Enhancing the Value and Sustainability of Field Stations and Marine Laboratories in the 21st Century

Committee on Value and Sustainability of Biological Field Stations,
Marine Laboratories, and Nature Reserves
in 21st Century Science, Education, and Public Outreach

Board on Life Sciences

Division on Earth and Life Studies

NATIONAL RESEARCH COUNCIL
OF THE NATIONAL ACADEMIES

THE NATIONAL ACADEMIES PRESS
Washington, D.C.
www.nap.edu

THE NATIONAL ACADEMIES PRESS • 500 Fifth Street, N.W. • Washington, DC 20001

NOTICE: The project that is the subject of this report was approved by the Governing Board of the National Research Council, whose members are drawn from the councils of the National Academy of Sciences, the National Academy of Engineering, and the Institute of Medicine. The members of the committee responsible for the report were chosen for their special competences and with regard for appropriate balance.

This report was supported by grants from the National Science Foundation (1012215).

Any opinions, findings, conclusions, or recommendations expressed in this publication are those of the authors and do not necessarily reflect the views of the organizations or agencies that provided support for the project.

International Standard Book Number-13: 978-0-309-30534-1
International Standard Book Number-10: 0-309-30534-9

Additional copies of this report are available for sale from the National Academies Press, 500 Fifth Street, N.W., Keck 360, Washington, DC 20001; (800) 624-6242 or (202) 334-3313; http://www.nap.edu/.

Printed in the United States of America

Front Cover:

Top Left: Students take measurements of a gray fox, Urocyon cinereoargenteus, before fitting it with a radio transmitter collar at Quail Ridge Reserve in Napa, California. Photo by Arielle Crews, UC Natural Reserve System.

Top Right: Scientists from the Mountain Studies Institute collect lake sediment cores at Crater Lake. Photo provided by Mountain Studies Institute (www.mountainstudies.org).

Middle Left: Scientist with Bluntnose Sixgill Shark on marine research vessel. Photo provided Florida State University Coastal and Marine Laboratory.

Middle Right: Scientists conduct a laboratory experiment at LUMCON. Photo provided by Nicole Cotten, Louisiana University Marine Consortium.

Bottom Right: Flathead Lake Biological Station (aerial view) is located on a peninsula of native forest on the east shore of Montana's Flathead Lake. Photo provided by Flathead Lake Biological Station.

Bottom Left: Front view of Therkildsen Field Station at Emiquon, University of Illinois Springfield. Photo by Melissa Benedict.

Back Cover:

Top Left: Installation of internet network equipment on Agave Hill tower for Boyd Deep Canyon Desert Research Center. Photo by Mark Fisher, UC Natural Reserve System.

Top Right: Botany class at Bodega Marine Reserve, Bodega Marine Lab in the background. Photo by Jackie Sones.

Middle Left: Scientists from the Flathead Lake Biological Station collect aquatic insects from a floodplain spring. Photo provided by Flathead Lake Biological Station.

Middle Right: Scientist at Hastings Natural History Reservation studying woodland star plants, Lithophragma sp. Photo by Mark Stromberg.

Bottom Left: Aerial view of Louisiana University Marine Consortium. Photo by Nicole Cotten.

Bottom Right: Lake Erie Center, University of Toledo. Photo by Donald Kemp.

THE NATIONAL ACADEMIES

Advisers to the Nation on Science, Engineering, and Medicine

The **National Academy of Sciences** is a private, nonprofit, self-perpetuating society of distinguished scholars engaged in scientific and engineering research, dedicated to the furtherance of science and technology and to their use for the general welfare. Upon the authority of the charter granted to it by the Congress in 1863, the Academy has a mandate that requires it to advise the federal government on scientific and technical matters. Dr. Ralph J. Cicerone is president of the National Academy of Sciences.

The **National Academy of Engineering** was established in 1964, under the charter of the National Academy of Sciences, as a parallel organization of outstanding engineers. It is autonomous in its administration and in the selection of its members, sharing with the National Academy of Sciences the responsibility for advising the federal government. The National Academy of Engineering also sponsors engineering programs aimed at meeting national needs, encourages education and research, and recognizes the superior achievements of engineers. Dr. C. D. Mote, Jr., is president of the National Academy of Engineering.

The **Institute of Medicine** was established in 1970 by the National Academy of Sciences to secure the services of eminent members of appropriate professions in the examination of policy matters pertaining to the health of the public. The Institute acts under the responsibility given to the National Academy of Sciences by its congressional charter to be an adviser to the federal government and, upon its own initiative, to identify issues of medical care, research, and education. Dr. Victor J. Dzau is president of the Institute of Medicine.

The **National Research Council** was organized by the National Academy of Sciences in 1916 to associate the broad community of science and technology with the Academy's purposes of furthering knowledge and advising the federal government. Functioning in accordance with general policies determined by the Academy, the Council has become the principal operating agency of both the National Academy of Sciences and the National Academy of Engineering in providing services to the government, the public, and the scientific and engineering communities. The Council is administered jointly by both Academies and the Institute of Medicine. Dr. Ralph J. Cicerone and Dr. C. D. Mote, Jr. are chair and vice chair, respectively, of the National Research Council.

www.national-academies.org

COMMITTEE ON VALUE AND SUSTAINABILITY OF BIOLOGICAL FIELD STATIONS, MARINE LABORATORIES, AND NATURE RESERVES IN 21ST CENTURY SCIENCE, EDUCATION, AND PUBLIC OUTREACH

JERRY R. SCHUBEL (*Chair*), Aquarium of the Pacific, Long Beach, California
FELICIA C. COLEMAN, Florida State University Coastal and Marine Laboratory, St. Teresa
CATHY CONRAD, Saint Mary's University, Halifax, Nova Scotia, Canada
DIANE DEBINSKI, Iowa State University, Ames
PETER M. KAREIVA, The Nature Conservancy, Seattle, Washington
GEORGE I. MATSUMOTO, Monterey Bay Aquarium Research Institute, Moss Landing, California
DIANE M. MCKNIGHT, University of Colorado Boulder
CAMILLE PARMESAN, Plymouth University, Plymouth, UK; University of Texas, Austin
ROBERT PLOWES, University of Texas Brackenridge Field Laboratory, Austin
ALISON G. POWER, Cornell University, Ithaca, New York
MARY E. POWER, Angelo Coast Range Reserve, University of California, Berkeley
MARK R. STROMBERG, University of California Natural Reserve System (retired), Sonoita, Arizona

STAFF
KEEGAN SAWYER, Study Director and Program Officer
CLAUDIA MENGELT, Senior Program Officer
LAURENCE YEUNG, Christine Mirzayan Science & Technology Policy Fellow (until April 2014)
LAUREN SONI, Senior Program Assistant
NORMAN GROSSBLATT, Senior Editor

OCEAN STUDIES BOARD

I believe that in the not too distant future a much larger share of biological research, from biochemistry to ecology, will be conducted at field stations that consist of nature preserves and have ready access to laboratories equipped to analyze and monitor processes at every level of biological organization, including the molecular. Field stations will also serve as key centers of education at all levels. Universities and other institutions wise enough to invest in such stations now, even in the face of limited financial resources, will assure themselves of a much larger share in the future action.

Edward O. Wilson

Field stations provide the best connection between a growing population and the wonders and mysteries of the natural environment. These institutions educate on what all citizens must do to preserve ocean health, the foundation of the basic ecosystem services that keep our planet habitable.

Marcia McNutt

Preface

The National Science Foundation (NSF) arranged for a review by the National Academy of Sciences to assess and explore mechanisms, in a time of declining resources, to maintain and enhance the important contributions of field stations, marine laboratories, and nature reserves in scientific discovery, innovation, education, and public outreach—roles encompassed by the missions of these institutions. In response, the National Research Council established the Committee on Value and Sustainability of Biological Field Stations, Marine Laboratories, and Nature Reserves in 21st Century Science, Education, and Public Outreach, which prepared this report. Biographic information on the committee members is presented in Appendix B.

In the course of preparing this report, the committee met three times in person and once by teleconference. During its deliberations, it heard oral presentations by the following: John Wingfield, Scott Edwards, Peter McCartney, Kandace Binkley, and David Campbell (NSF); Guy Noll, Morakot Pilouk, Marten Hogeweg, and Jeff Donze (Esri); Hillary Swain (Archbold Biological Station); Ian Billick (Rocky Mountain Biological Laboratory); Ivar Babb (University of Connecticut Northeast Underwater Research, Technology & Education Center); Clarissa Dirks (Evergreen State University); Diane Ebert-May (Michigan State University); Caroline Wagner (Ohio State University); and Anthony Michaels (Proteus Environmental Technologies). Interested members of the public at large were given an opportunity to speak at the first meeting. In addition to the information from those presentations and the peer-reviewed scientific literature, the committee made use of field station databases provided by the National Association of Marine Laboratories, the Organization of Biological Field Stations, and the National Geographic Society. The committee acknowledges and thanks those individuals and groups for their valuable input.

This report has been reviewed in draft form by persons chosen for their diverse perspectives and technical expertise in accordance with procedures approved by the National Research Council Report Review Committee. The purposes of the review are to provide candid and insightful comments that will assist the institution in making the published summary as sound as possible and to ensure that the summary meets institutional standards of objectivity, evidence, and responsiveness to the study charge. The review comments and draft manuscript remain confidential to protect the integrity of the deliberative process. We thank the following individuals for their review of this report:

George Crozier, Daulphin Island Sea Lab
William Farland, Colorado State University
Elisabeth Gantt, University of Maryland
Gary Jacobs, Strata-G LLC
Geraldine Knatz, Bank of the West
Terry McGlynn, California State University, Dominguez Hills

Holly Menninger, North Carolina State University
Dwayne E. Porter, University of South Carolina
Shawn Rowe, Oregon State University
Joshua Tewksbury, University of Washington
Henry M. Wilbur, University of Virginia

Although the reviewers listed above provided many constructive comments and suggestions, they were not asked to endorse, nor did they see, the final version of the report before its release. The review of the report was overseen by May R. Berenbaum of the University of Illinois at Urbana-Champaign and John E. Burris of the Burroughs Wellcome Fund. Appointed by the National Academies, they were responsible for making certain that an independent examination of the report was carried out in accordance with institutional procedures and that all review comments were carefully considered. Responsibility for the final content of the report rests entirely with the authoring committee and the National Research Council.

The committee's work was assisted by the staff of the National Research Council's Board on Life Sciences and Ocean Studies Board. We thank the study directors, Keegan Sawyer and Claudia Mengelt. The world map of field stations was made possible through the research support of Laurence Yeung and Sarah Gizaw. We also acknowledge and thank Rob Greenway, Sharon Martin, Mirsada Karolic-Loncarevic, Stacee Karras, Carl Anderson, Sayeeda Ahmed, Payton Kulina, and Lauren Soni for their technical and logistical support.

Jerry R. Schubel, Chair

Committee on Value and Sustainability of Biological Field Stations, Marine Laboratories, and Nature Reserves in 21st Century Science, Education, and Public Outreach

Contents

Tables

Summary

Recognizing the value of field stations, marine laboratories, and nature reserves for research, education, and public outreach and in light of their current challenges, the National Science Foundation (NSF) asked the National Research Council to address the following tasks: to summarize—on the basis of previous reports—field stations' value to science, education, and outreach; to outline strategies to meet future research, education, outreach, infrastructure, and logistical needs of field stations; to explore ways in which field stations could network more broadly; to evaluate field stations' contributions to research, innovation, and education; and to suggest long-term financial strategies to sustain field stations' missions (see Statement of Task in Appendix A).

For over a century, field stations[1] have been important entryways for scientists to study and make important discoveries about the natural world. They are centers of research, conservation, education, and public outreach often embedded in natural environments that range from remote to densely populated urban locations. Field stations vary greatly in size and sophistication of infrastructure. Long-term research at field stations produces baseline and sentinel data that can be used to study ecosystems at a time when human activities are altering nature at an unprecedented rate.

Most field stations are affiliated with universities.[2] Because they lack traditional departmental boundaries, researchers at field stations have the opportunity to converge their science disciplines in ways that can change careers and entire fields of inquiry. Field stations provide physical space for immersive research, hands-on learning, and new collaborations that are otherwise hard to achieve in the everyday bustle of research and teaching lives on campus. But the separation from university campuses that allows creativity to flourish also creates challenges. Sometimes, field stations are viewed as remote outposts and are overlooked because they tend to be away from population centers and their home institutions. This view is exacerbated by the lack of empirical evidence that can be used to demonstrate their value to science and society. Today's technologies—such as streaming data, remote sensing, robot-driven monitoring, automated DNA sequencing, and nanoparticle environmental sensors—provide means for field stations to retain their special connection to nature and still interact with the rest of the world in ways that can fuel breakthroughs in the environmental, physical, natural, and social sciences. The intellectual and natural capital of today's field stations present a solid platform, but many need enhancements of infrastructure and dynamic leadership if they are to meet the challenges of the complex

[1]In this report, for the sake of brevity, the committee refers to field stations, marine laboratories, and nature reserves as field stations.

[2]Seventy-five percent of field stations are university affiliated according to a 2012 survey conducted by the Organization of Biological Field Stations and the National Association of Marine Laboratories (NAML-OBFS 2013b).

problems facing the world. This report focuses on the capability of field stations to address societal needs today and in the future.

Science for an Unpredictable World

The rapid environmental changes that are taking place globally raise basic research questions and present major societal challenges. Evidence is mounting that the growing human footprint is stressing natural and social systems. Climate change, biodiversity loss, natural resource extraction, and pollution pose considerable threats to ecosystems, economies, and human well-being. Coping with the challenges will require improved knowledge about the social–ecological system. Field observations have played and will continue to play critical roles in the physical, natural, and social sciences.

Field stations are national assets formed by the unique merger of natural capital, intellectual capital, social fabric, and infrastructure that leads to the important scientific endeavors required if we are to understand our rapidly changing natural world. Field stations, either inadvertently or by design, are repositories of long-term observations and datasets of natural history necessary for documenting global changes. A greater emphasis on integrated, multidisciplinary research that includes the physical sciences, geosciences, social sciences, humanities, and arts will enhance scientists' use of historical datasets to address global challenges. The recognition of the importance of this portfolio of activities in what is now called "convergence"[3] is a strength of many field stations.

> ***Recommendations***: Field station leaders should identify and support the development of scientific and educational assets that harness their stations' unique qualities to address local, regional, national, and global challenges by bringing together scientists from a number of disciplines, including the social sciences, through what is now called convergence.

Preparing Our Next Generation of Scientists

Recruiting students into fields of science, technology, engineering, and mathematics (STEM) has been identified as having high priority in many nations, given the importance of STEM fields for innovation and economic growth. Field stations are venues for discovery-based learning,[4] and they offer rich opportunities for other types of active learning, which have been shown to promote diversity and

---

[3] Convergence is an approach to problem solving that cuts across disciplinary boundaries. It integrates knowledge, tools, and ways of thinking from life and health sciences, physical, mathematical, and computational sciences, engineering disciplines, and beyond to form a comprehensive synthetic framework for tackling scientific and societal challenges that exist at the interfaces of multiple fields (NRC 2014a)

[4]Discovery-based learning, also called inquiry-based learning, requires students to pose their own questions and develop hypotheses and to design experiments to address their questions (Johnson and Lawson 1998). It is a type of active learning, a student-centered approach to instruction, which requires students to engage in meaningful learning activities (Dirks 2011).

persistence in STEM fields. Integration of research into formal and informal education and into public outreach activities provides engaging learning opportunities for people of all ages and backgrounds.

> ***Recommendation***: Universities and other host institutions should expand opportunities at field stations to conduct independent and collaborative research and active learning activities to increase interest and persistence in STEM fields.

Empowerment Through Engagement

Public understanding and participation in science is important in increasing human connectedness to the natural world and empowers citizen decision making and involvement in public policy. Field stations support a wide range of public outreach and engagement programs—public lectures and workshops, science cafes, field trips, and nature walks, among other informal education opportunities—to enhance public understanding of science. The committee applauds these public outreach efforts because they break the mold of traditional science communication and more actively involve public audiences in science. However, field station engagement programs are often disconnected from empirically based approaches to develop and evaluate effective science communication and informal education activities.

> ***Recommendation:*** Field stations should continue to explore a wide range of approaches to engage the public in science, and select and tailor their activities in a manner that best leverages a field station's location, personnel, infrastructure, and other available resources. Empirically based approaches in science communication and informal education should be used to guide the development and assessment of engagement activities to promote public understanding of science effectively.

Citizen science is an emerging channel through which field stations can advance science and empower people interested in science by engaging them actively in data collection and research, particularly in science issues that affect their communities. There is a broad spectrum of citizen science initiatives, from simple observational programs to coordinated, training-intensive environmental monitoring programs. Citizen science initiatives enable people to learn about science and the ecosystem dynamics of natural communities in which their field stations are embedded. Citizen science initiatives also can enable coordinated networks of volunteers to collect data that can inform our understanding of how human activities may be altering ecosystems. Much of citizen science is facilitated through advances in Web-based technologies that allow citizens to collect and analyze data through accessible platforms, such as smart phones and personal computers. A few field stations have developed sustained outreach programs that include citizen science, but citizen science initiatives are not yet widespread among field stations.

Recommendation: Field stations should collaborate in, connect with, and formalize citizen science programs by using the latest technologies and networking initiatives throughout the U.S. and global system of stations and thus offer a coordinated infrastructure for interested members of the public to engage in, learn about, and contribute to science.

Networking for Discovery and Innovation

Most field stations operate independently of one another. Greater networking with other field stations and with research centers would be beneficial because it could leverage resources to facilitate discovery and spark innovation. Networking would also allow field stations to share best practices, protocols, and platforms for data archiving and retrieval. Such networking has the potential to open new arenas of scientific inquiry, education, and outreach. It can capture social and intellectual capital to tackle major questions and seize opportunities as no single field station can, and it enhances creativity and innovation by attracting a wide range of scientists and promoting multidisciplinary collaboration. The most successful and sustainable networks start small and are self-defining; they encourage reciprocity among network members. Networking can facilitate the development and diffusion of knowledge and technology in a way that encourages innovations.

Recommendation: Field stations should seek opportunities for networking that make scientific, educational, and business sense. Universities and funding organizations should provide incentives for networking of field stations that meet those criteria. NSF and other funding agencies could encourage networking of field stations through the request-for-proposal process by giving preference to proposals that link multiple field stations.

Modern Infrastructure for a Networked World

Field stations vary in scope, size, and purpose; each contributes to the global portfolio in distinct ways. There is no single array of infrastructure that is applicable to all field stations, although there are some similar needs across field stations of differing sizes and complexity. Internet connectivity and cyberinfrastructure[5] are two neglected and underdeveloped elements of field station infrastructure. Adequate Internet connectivity and cyberinfrastructure would facilitate the task of bringing dark data[6] to light, extending the range of accessible natural history, and would improve networking for discovery. Installation of new cyberinfrastructure requires data-management and data-sharing plans and conformity of data with widely used metadata standards. Such infrastructure also requires a long-term

---

[5]Cyberinfrastructure refers to the assortment of information technologies that enable data storage, management, integration, and analysis.

[6] Data that are not systematically indexed or stored in a manner that is accessible to the broader scientific community, such as biological specimen collections, analogue data (e.g., observations recorded in laboratory notebooks), and data found only in research publications (Heidorn 2008).

funding commitment for repair, upgrades, and technical support.

> ***Recommendation***: Because of their wide variety in purpose, size, and scope, each field station should assess and define its own infrastructure needs. However, Internet connectivity and cyberinfrastructure should be included in all infrastructure-management plans to allow field stations to facilitate collaborative research and participate in broader networking efforts. The process of archiving dark data into digitally accessible formats is critical, and should begin with the most recent datasets and progress back in time so that field stations can expand their sets of continuous longitudinal data.

Financial Security for a Modern Infrastructure

Aging infrastructure, the need for advanced technology and cyberinfrastructure, and evolving safety regulations are increasing financial demands on field stations as they upgrade to meet emerging science and societal challenges. Sustainable funding for modern infrastructure will be possible only if field station leaders develop compelling value propositions, strategic plans, and business models for operations that can secure base funding support that in turn can be leveraged by support from diverse sources. However, field station leaders too often lack entrepreneurial skills. Effective business planning requires strong linkages to funding institutions and reaching out to diverse constituencies that can derive value from field stations.

> ***Recommendation***: Field stations and their host institutions should develop business plans that include clear value propositions and mechanisms to establish reliable base funding commitments that can be supplemented with funding from diverse sources. Business planning requires that station leaders be recruited not only for their scientific credentials, but also for their leadership, management, and entrepreneurial skills. Host institutions should provide mentoring of field station leaders in management, business planning, and fundraising when appropriate.

Measuring Performance and Impact

The value of field stations is widely but unevenly documented by scientists in anecdotal evidence and in qualitative and semi-quantitative data. Measures of effectiveness tfor example, the number of archived digital datasets, the number of students conducting independent research projects, and the award amounts of grants—that are aligned with a host institution's science, education, and business plans can lead to improvement in performance and impact but these typically are lacking.

In the absence of metrics, it is impossible to manage for improved outcomes. Field stations would benefit from consistent, comparable metrics to modify, monitor, and assess their strategies for meeting goals in research, innovation, education, training, outreach, and engagement. Discovering, sharing, and archiving such metrics from field stations are critical. The development of digital object identifiers for field stations is a potential starting point for collecting data that can be transformed into metrics and information. Metrics for quantifying the value of field stations to science and society are essential if field stations are to be justified to supporters.

> ***Recommendation***: Field stations should work together to develop a common set of metrics of performance and impact. The metrics should be designed so that they can be aggregated for regions and the entire nation. Universities and other host institutions and funding organizations should support the gathering and transparent reporting of field station performance metrics because such information will enhance the stations' ability to document their contributions to the nation's research and education enterprise.

> ***Recommendation***: New mechanisms and funding need to be developed to collect, aggregate, and synthesize performance data for field stations, and to translate these data into metrics and information that can be used to document the value of the community of field stations to science and society.

1
Contributing to Science and Society

> *The voyage of the "Beagle" has been by far the most important event in my life, and has determined my whole career. . . . I have always felt that I owe to the voyage the first real training or education of my mind; I was led to attend closely to several branches of natural history, and thus my powers of observation were improved, though they were always fairly developed.*
>
> —Charles Darwin, 1887

Captain Robert FitzRoy unwittingly altered the course of the scientific enterprise when by serendipity he engaged a young naturalist to join him on a 5-year sea voyage to Tierra del Fuego. Fitzroy, a gifted meteorologist and a career officer in the Royal Navy, came to the helm of his vessel, HMS *Beagle,* quite unexpectedly. The loneliness of the sea, it seemed, had led the ship's previous captain to take his life during the *Beagle's* first research voyage to South America. To guard against a similar fate, FitzRoy requested the accompaniment of a science-minded companion to keep him engaged—Charles Darwin.

On December 27, 1831, Fitzroy and Darwin set sail on their famous voyage aboard HMS *Beagle.* Fitzroy provided a crucible—a mobile biological field station—that gave Darwin unprecedented access to pristine natural environments, where he recorded careful observations on geology as well as the behavior, physical shape, and ecology of plants and animals. The result was that Darwin provided the world with the key to modern biology: the theory of natural selection.

Naturalists such as Darwin who observed and described the world around them laid the foundation for such scientific disciplines as biology, physics, and biogeography (Dolan 2007, Wyman et al. 2009). Many of their observations were made from field camps and stations, marine laboratories, and nature reserves, all referred to herein for brevity as field stations. Field stations have long been stewards of place-based historical data on our natural world. In this report, we use the definition of a field station in Box 1-1, and this definition includes marine laboratories and natural reserves.

BOX 1-1
Definition of a Biological Field Station, Marine Laboratory, or Nature Reserve[a]

A field station is a center of scientific research, conservation, education, and outreach that is embedded in the environment in a location that is usually protected and that serves both the local community and the larger scientific community. The research conducted at a field station is often focused on local environmental regions, but national and international scientific projects are common.

[a]Referred to herein for brevity as field stations.

Study Approach

The National Science Foundation (NSF) recognizes the values and vulnerabilities of field stations and welcomes guidance on positioning them to advance science and society in a financially sustainable manner. To that end, NSF asked the National Research Council to review and assess the roles that field stations play in promoting and supporting research in science and engineering, in education at all levels, and in outreach to policy makers and the general public (see the Statement of Task in Appendix A). NSF is also interested in investigating new modes of operation of field stations that include enhanced engagement of scientists in different countries and of citizen scientists, and in nurturing closer ties with their local communities. NSF asked the National Research Council to give special consideration to collaborative mechanisms through which field stations can work with one another, nationally and internationally, and with state and federal research facilities to enhance their research and training programs and to reduce duplicative efforts.

In responding to the Statement of Task, the committee encountered a significant challenge to empirically demonstrating the value of field stations due to the lack of aggregated data on their activities and impacts on science and society. Some field stations collect data about their individual programmatic impacts, although the data may not be publically available. No recent attempts have been made to aggregate data across the community of field stations such as trends in station activities, contributions to research publications or public policy reports, programmatic outcomes and impacts, or other data that could be used to enumerate how field stations over time have contributed to science, education, and public outreach. Quantitative measures of the current status of field stations and historical trends in field station use and support would likely boost arguments for a broad investment in the enterprise. For example, annual data on the use of field stations by researchers, students, and the public could indicate trends in demand for this infrastructure. In the absence of this information, unless otherwise noted, the committee relied on its collective experience, publicly available data on individual and small networks of field stations, and anecdotal evidence to characterize the community of field stations and their value to science and society.

What Is a Field Station?

Field stations constitute critical infrastructure for the scientific enterprise. More than 900 field stations are scattered around the world (Figure 1-1). Field stations vary greatly in size, sophistication of infrastructure, and distance from population centers. For example, a field station may be a rustic shelter within a gated or fenced-in nature reserve or a sophisticated marine laboratory with modern research equipment and vessels, laboratory space, residential housing, and conference facilities. On both ends of the spectrum, they provide environments to observe nature where access is relatively controlled and experimental setups are relatively protected from tampering. A few publications provide basic information about an aggregate of field stations, although none of these assess the impacts or value of field stations to science and society (Table 1-1).

The most recent and comprehensive survey found that approximately 75 percent of field stations that hold U.S. mailing addresses are overseen by a college or university (NAML-OBFS 2013b). Informal studies suggest that fewer overseas stations are university affiliated (Dolan 2007, Wyman et al. 2009), but formal assessments of field stations around the globe have not been conducted to confirm this finding.

Many field stations are affiliated with the Organization of Biological Field Stations (OBFS), the National Association of Marine Laboratories (NAML), or their international counterpart. OBFS supports its members by developing relationships with funders, cooperating in research networks, and sharing information with representatives in Washington D.C., but it does not serve as a central administration. Similarly, NAML's mission is to stimulate research and promote education while providing its membership with strong public policy support and a venue for resolving problems common to most nonprofit marine laboratories in the nation. Neither OBFS nor NAML is considered a formal network or provides a management structure for its members. However, OBFS and NAML play important organizing roles for the field station community discussions about the value of field stations to science and society, and approaches to prepare field stations for the future.

TABLE 1-1 Aggregated Information About Field Stations from Three Publications

Publication	Data Gathering	*N*[a]	Geography	Annual Operating Budget	Primary Affiliations
NAML-OBFS (2013b)	Formal survey in 2012	197-218	Field stations and marine labs with U.S. mailing addresses	16.8%, <$50k 26.9%, $50k-$250k 47.2%, $250k-$5M 9.1%, >$5M	74% University 14% Government 11% NGO[b] 7% Other (*n*=202)
Whitesell et al. (2002)	Formal survey in 1997	66	tropical biological field stations (33 countries)	$846-$2.9M Avg = $323,811 Median = $85k	Not reported
Wyman et al. (2009)	Informal questionnaire 1993-2007	90-201	International field stations, marine labs, and agricultural stations[c]	Not reported	35.4% University 34.3% Government 26.1% NGO 4.1% Other

[a] Range provided when not all respondents answered every survey questions.
[b] NGO = nongovernmental organization.
[c] In many countries overseas, agricultural research stations are available to conduct ecological research or natural history studies.

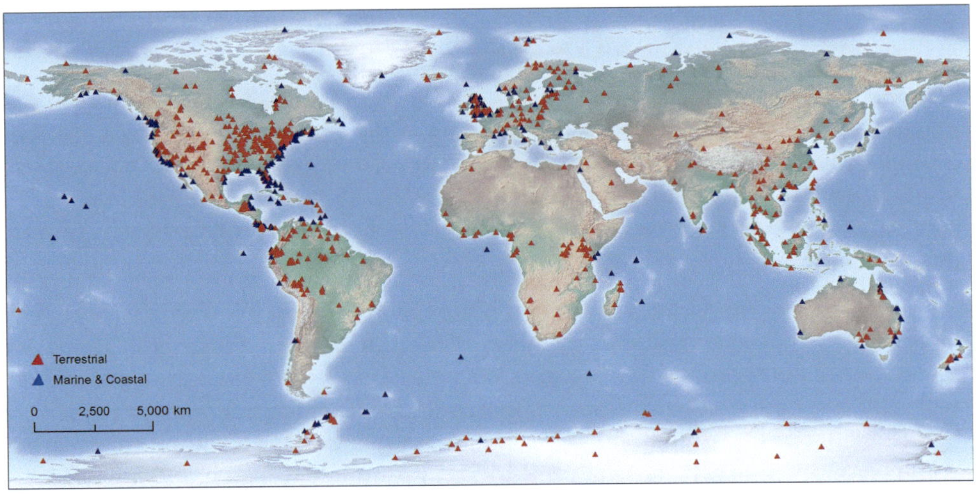

FIGURE 1-1 World map of biological field stations and marine laboratories. The global distribution of 963 terrestrial, coastal, and marine stations for which current operating status and geographic location could be determined. Information for approximately two-thirds of the stations was determined from databases provided by the National Association of Marine Laboratories, the Organization of Biological Field Stations, and the Royal Geographical Society. Station information was also obtained from websites of the Association of European Marine Biological Laboratories, the Canadian Society for Ecology and Evolution, the Chinese Ecosystem Research Network, the Institute of Biological Problems of the North, the Japanese Association of Marine Biology, the International Network for Terrestrial Research and Monitoring of the Arctic, the Smithsonian Tropical Research Institutes, the Tropical Ecology Assessment and Monitoring Network, the World Association of Marine Stations, and the Google search engine.

Enabling Scientific Discovery

Long-term datasets, coupled with monitoring and experimentation, provide mounting evidence that the human footprint is "stressing natural and social systems beyond their capacities" (Millennium Ecosystem Assessment 2005, IPCC 2007, NSF 2009). Indeed, forecasts for the remainder of the 21st century suggest that Earth will undergo global changes at an increasing rate (NRC 2010a,b 2013; AAAS 2014). Climate change, overexploitation and pollution of natural resources, and instabilities in food production pose considerable threats to ecosystem resilience and to the mental, physical, and economic health of people and nations. These stressors present major societal challenges for which substantial data and infrastructure are needed (EPA 2012).

Research conducted at field stations enhances scientists' ability to make reliable, robust projections of change that can help decision makers identify, evaluate, and choose among potential actions. For example, the relatively undisturbed conditions that exist at many field stations combined with long-term data on populations, communities, and baseline environmental conditions make these sites fruitful for assessing climate change impact. Long-term data on

TABLE 1-2 Comparison of Four Global Metal-Analyses of the Impact of Climate Change on Wild Species[a]

Study[b]	*N*: Species + Functional Groupings	% Changing Distribution + Phenology	% Change Consistent with Climate Change	*p*-Value[c]	% of Studies Conducted at Terrestrial Field Stations
Parmesan and Yohe (2003)	1598	59	84	$<10^{\,13}$	28
Root et al. (2003)	1468	40	82.3	$<10^{\,13}$	31
Parmesan (2007)	202	78	91	$<10^{\,7}$ [d]	42
Rosenzweig et al. (2008)	Not specified	_	90	<0.001	33

[a] Data presented reflect only terrestrial studies. Publications on impact of climate change on marine life were excluded because of the difficulty of assessing the extent to which marine laboratories facilitated the research (i.e., information was not provided in the publications)
[b] These four meta-analyses publications are heavily cited in the scientific literature (nearly 8,000 total citations in Google Scholar as of March 30, 2014) and contributed substantially to the Third, Fourth, and Fifth Intergovernmental Panel on Climate Change Assessment Reports (IPCC 2001, 2007, 2014)
[c] Binomial probability for the percent change that is consistent or inconsistent with local and regional climate change.
[d] *p*-value calculated from original dataset, but not provided in the publication.

phenological events (e.g., yearly dates of bird breeding, leaf budding, butterfly emergence, arrival of migratory species), population dynamics, or even species presence and absence can be analyzed for long-term trends and linked to trends in local or regional climate. Field stations figure prominently in a number of major global meta-analyses of the impacts of anthropogenic climate change on the distributions of wild species, accounting for 28-43 percent of the studies included in the analyses (Table 1-2). This body of research clearly has shaped international greenhouse gas policies, as evidenced by its consistently high profile in the assessment reports of the Intergovernmental Panel on Climate Change (IPCC 2001, 2007, 2014). In particular, this work has been crucial for assessing "dangerous" levels of anthropogenic contributions to climate change (Hansen et al. 2013), and contributed to the international agreement to keep global warming below a 2°C threshold (UNFCCC 2009).

Field stations enable scientists to discover and increase knowledge about biological and physical processes that govern our world and to document, forecast, and design strategies to adapt to and mitigate a wide array of environmental and ecosystem challenges. They can be thought of as nodes in a sensing network to monitor changes in the environment. The long-term observations across a range of landscapes—from the relatively pristine to urban or agricultural areas—form important and irreplaceable historical records of environmental changes and the impact of human activities. For example, field stations have supported discoveries ranging from the interconnectedness of food webs to the geographic patterns of the spread of disease to the extent and consequences of global climate change—discoveries that required long-term, place-based research (Michener et al. 2009, Billick and Price 2010). Field stations provide windows into ecosystems that may not be otherwise readily available to scientists (Box 1-2). The longitudinal baseline and time-series data collected at field stations can be used to evaluate environmental change and the forces that drive it. Field station data have proven to be vital for forecasting future change (Billick et al. 2013). As a result, the landscapes surrounding field stations often are intensively studied ecosystems "in which the steady accumulation of site-specific knowledge becomes a powerful platform for future research" (Billick et al. 2013).

Field stations make up a critically important component of the nation's research capital that is complemented by protected lands in parks and forests, in land conservation trusts, and on private property. The infrastructure of field stations offers unique advantages to research in terms of place-based logistical support and equipment. Field stations support continued access to protected study sites and relatively secure placement of conspicuous experimental materials (cages, markers, and other equipment) that enable scientists to collect long-term data to document local natural history. In addition, field stations have inspired countless young people to pursue careers in science and have trained countless more. Field stations are important for science and education in a world that is changing at an unprecedented rate, and their value to society only grows with time.

Education, Outreach, and the Building of a Scientific Community

Field stations are important for STEM education and training at all levels. Many young people have been drawn to science—whether to pursue it as a career or avocation or to advocate for scientific endeavors—because of a visit to a field station. They include the members of the current committee, who on average had their first field station experiences 36 years ago. That fact helps to validate the integration of scientific research with formal and informal education as an important endeavor for field stations, and it should encourage field researchers to find opportunities to engage students and other citizens as part of their research teams, offering them hands-on research experiences (Billick et al. 2013). Moreover, students often contribute substantially to research conducted at field stations (Box 1-3), advancing science as they learn.

In a recent survey conducted by the NAML and the OBFS, more than 90 percent of the 227 respondents reported that their field stations serve academic research scientists, graduate students, and undergraduates (NAML-OBFS 2013b). Numerous Research Experiences for Undergraduates at field stations are either supported directly by the stations or by government programs (NSF 2013a, b).[7]

---

[7]See NSF list of Research Experiences for Undergraduates sites, http://www.nsf.gov/crssprgm/reu/list_result.jsp, and NSF Experimental Program to Stimulate Competitive Research (EPSCOR), http://www.nsf.gov/od/iia/programs/epscor/nsf_oiia_epscor_index.jsp.

Many field stations also have postgraduate research students and postdoctoral fellows on site who participate in research, teaching, and outreach activities with local communities.

Field experiences are ideal for discovery-based learning,[8] which can help improve a student's science scores, self-esteem, conflict resolution, problem solving, motivation to learn, and classroom behavior (American Institutes for Research 2005). Field stations draw learners of all ages into hands-on learning in real-world classrooms. These learners often differ from those found on university campuses in that they might include elementary school students on field trips, city council members participating in seminars on enhancing community resilience, a university provost who explores options for campus green building initiatives, or a senator who wants to understand the nuances of a state's ecosystem-health report card. A growing and more sustained presence at field stations includes members of the general public who are participating in research initiatives.

Field stations facilitate learning—by citizens of all ages, from kindergarten age to adulthood—about local natural history and engagement in science. More than 60 percent of field stations serve K-12 students, the general public, and state or federal government scientists (NAML-OBFS 2013b). Many outreach programs at field stations focus on informal education through public lectures, workshops, science cafés, field trips, nature walks, and volunteer opportunities. These activities held at field stations and in nearby communities provide opportunities for the exchange of ideas between scientific staff at field stations and lay audiences. For example, the Nantucket Field Station of the University of Massachusetts Boston maintains an array of K-12 activities that include a Junior Ranger program for middle-school–age children and science internship programs for high school students. It also operates a volunteer program in which interested citizens can assist with maintenance, administrative tasks, and research. The University of Maryland's Chesapeake Biological Laboratory includes a volunteer-run visitor center and hosts free "science for nonscientists" outreach seminars. Some field station staff also contribute to outreach by advising decision makers ranging from civic groups to state governments. The University of Wisconsin Milwaukee Field Station staff provides advice to local community groups and state and federal agencies about natural history, conservation, and other issues associated with natural areas.

One way in which some field stations enhance public outreach is through citizen science programs (see Box 1-4). Citizen science provides a way for people interested in science to engage actively in understanding environmental issues that affect their communities and in supplementing and sustaining data streams that have been interrupted or curtailed by reductions in government funding for monitoring (Conrad and Hilchey 2011), and in developing new data streams not previously available.

---

[8]Discovery-based learning, also called inquiry-based learning, requires students to pose their own questions, develop hypotheses, and design experiments to address their questions (Johnson and Lawson 1998).

BOX 1-2
Invasive Fire Ants: The Hidden Value of Unwanted Guests

Red imported fire ants and a Phorid fly. Photo Credit: John & Kendra Abbott/Abbott Nature Photography.

In 1981, red imported fire ants (*Solenopsis invicta*) invaded the Brackenridge Field Laboratory (BFL) in Austin, Texas, and triggered a cascade of scientific inquiry that has expanded into a long-term research program on effects and biological control of invasive species. Fire ants, native to South America, are invasive pests in the United States, Australia, the Caribbean, and some eastern Asian countries. The United States spends an estimated $8 billion each year on fire ant control, damage mitigation, and medical treatment. Scientists at BFL have conducted extensive natural-history research, and the red ant invasion provided an opportunity to collect additional baseline natural-history data and track the effects of these invasive ants on the native arthropod community. Undergraduate students conducted some of the key initial studies of the ant invasion. A graduate student's work that documented how parasitoid (phorid) flies disrupt the foraging activities of a different ant species at BFL led directly to a major national program that uses phorid flies as biological control agents for fire ants (Feener 1981). Today, the BFL research group forms a key hub in the international fire ant research network, which includes collaborations that span continents. BFL maintains partnerships with more than 100 private landowners and agencies for region-wide evaluation studies and has established teaching and outreach programs about the challenges posed by invasive species in natural settings. The fire ant study has resulted in more than $10 million in research funds over 20 years, more than 80 publications, and a broader expert research program on invasive species.

BOX 1-3
Advancing Science and Education at Hopkins Marine Station

Left, Willis Hewatt at Hopkins Marine Station, 1935; Right, Raphael Sagarin at Hopkins Marine Station, 1994. Photo Credit: Hopkins Marine Station of Stanford University.

The oldest marine station on the West Coast of North America, Hopkins Marine Station, opened in 1892 as the Hopkins Seaside Laboratory and became the Marine Biological Laboratory of the Leland Stanford Junior University in 1906. In 1917, the field station moved to its current location and was renamed the Hopkins Marine Station of Stanford University (CENS_2013). The State of California designated the Hopkins Marine Life Refuge in 1931, and a graduate student, W.G. Hewatt, established a permanent 98.8-m-long intertidal transect (Hewatt 1934, 1937); marking the transect with brass bolts. Some 62 years later, two undergraduate students started a class research project to replicate Hewatt's research thesis. They produced one of the first studies to show that climate change was transforming a regional ecosystem (Barry et al. 1995, Sagarin et al. 1999) and demonstrate that students are often the driving force behind important science discoveries. Their work clearly demonstrated the importance of field stations for maintaining a protected environment and a long-term historical record of place-based research, and spurred a wave of similar research focused on species distributions.

Citizen science is not a new concept. The Audubon Christmas Bird Count[9], initiated over a century ago, demonstrates that citizens who have a passion for the environment and natural history can be counted on to deliver accurate data on species distributions and abundances of birds. By 1990, such activities by science-

[9]Audubon Christmas Bird Count website: http://birds.audubon.org/christmas-bird-count.

interested members of the public had become more formally known as citizen science. Today, citizen involvement in science is widespread, and volunteers are collecting valuable data that contribute to our understanding of ecosystems and of how human activities may be altering them. The spectrum of citizen science initiatives is broad, from relatively simple observational programs—such as iNaturalist, eBird, the Reef Environmental Education Foundation programs, and National Geographic's Bioblitz—to coordinated, training-intensive water-quality monitoring programs (Bowser and Shanley 2013). Some of the programs are coordinated by field station scientists or conducted at field stations themselves. Those citizen science initiatives help to meet conservation goals and empower citizens to engage in the science that is essential for solving ecological and economic problems that result from overexploitation of natural resources, loss of critical habitat, and unexpected and catastrophic events.

What draws learners and researchers of all ages to field stations? Foremost are the ecosystems within which the stations are embedded. Field stations typically are near or embedded in relatively pristine environments. They provide researchers and students with access to study areas that offer some level of protection for scientists working alone in remote areas and protection of their equipment from vandalism and unintentional damage by visitors. The natural history often is well documented and featured prominently in the scientific literature, particularly for field stations that have supported scientific endeavors for long periods. Other qualities that draw people to field stations include the infrastructure that facilitates research (e.g., housing, library, herbarium collections, Internet access, laboratory space and equipment, historical data, and personnel) and the sense of community. The "station culture" that thrives in field stations creates rich opportunities for students and faculty (including artists, engineers, life scientists, and social scientists) to form new collaborations and friendships that lead to broad discussions and often to serendipitous scientific discoveries (Michener et al. 2009). Many field stations engender communities to which people return year after year to share knowledge, their concern for one another, and their concern for the natural world. These shared experiences enable researchers and students at field stations to have free and uninhibited exchange of ideas. Resident staff members are important members of that culture. They often are devoted to the station mission, are engaged in the research, and serve as a station's ambassadors to the outside world through the researchers, teachers, students, and members of the public with whom they interact.

Field Stations in Jeopardy

Many field stations are in jeopardy. The lack of widespread recognition of their contributions to science and society leads to their systemic underuse and underfunding. Moreover, their often remote locations, low overhead support, and varied affiliations can result in disparate research networks of field stations that have inconsistent operational and organizational cohesion. In difficult budgetary environments, field stations—especially remote or small ones—are vulnerable to budget cuts and even closure. The vulnerability can be seen around the world. In

BOX 1-4
Citizen Scientists Contribute to Research on Global Warming

American pika (*Ochotona princeps*). Photo Credit: Sally King, U.S. National Park Service; http://www.nps.gov/band/naturescience/pika.htm.

The American pika (*Ochotona princeps*) lives in the alpine tundra of the Rocky Mountains. It is intolerant of high temperatures, so it is a potential sentinel of global warming. Chris Ray, of the University of Colorado, has been conducting research in the university's Mountain Research Station on the population ecology of the pika. She and researchers in the Colorado Division of Wildlife have partnered with local and regional organizations (Rocky Mountain Wild and the Denver Zoo) to support a citizen science effort to document the current distribution of pika. Their support includes training and the design of the observational program. Another partner in the effort is the Natural Resource Ecology Laboratory of Colorado State University, which hosts and manages the website through which Pika Patrol volunteers can upload their observations (http://www.adventureandscience.org/pika.html). Through those combined efforts, over 189 observations have been recorded since the effort began in 2011.

2008, repair costs and other budget concerns led the University of Hawaii at Manoa (UH Manoa) to announce closure of the 35-year-old Kewalo Marine Laboratory of the Pacific Biosciences Research Center (PBRC). The announcement caused protests from the UH Manoa faculty and the marine biology community. The battle ended 4 years later in November 2012, when an interim vice chancellor and a relatively new chancellor reversed the decision, allowing PBRC to apply for new research grants, search for new tenure-track faculty members, and begin a strategy for extending PBRC's K-12 science, technology, engineering, and mathematics (STEM) education programs (Pennisi 2008, Cruz 2012, Kalani 2012).

In a similar situation, the 45-year-old Experimental Lakes Area (ELA) freshwater research station in northern Ontario, Canada—renowned for its research on and monitoring of the effects of mercury, acid rain, and other contaminants on Canada's waterways—was scheduled to close in mid-2013. Fisheries and Oceans Canada chose to eliminate the ELA program after government budget cuts in 2012 (Orihel and Schindler 2014). However, the decision to close the ELA research station was reversed because of pressure from the local community and from academic and government scientists working at ELA (Hoag 2013). The outcry led to ELA's operational support being transferred from the federal Canadian government to the provincial Ontario government and the International Institute for Sustainable Development, a nonprofit research institute based in Canada (CBC News 2014). Also, in 2013, the University of London decided to close the University Marine Biological Station Millport in Scotland, which had been a crucial part of a network of research stations around the UK and European coast for over 100 years. Pressure from online petitions, social media, and organized campaigns led to a change in ownership, and the station is expected to reopen as the Millport Field Center in 2014 (BBC 2013).

Other field stations have not been as fortunate. In a historic review of biological field stations around the world, the most common reasons cited for closure included death of the founder or director, natural disaster, war, and curtailment of funds (Jack 1945). In recent cases, limited funding to support operations appears to be the primary reason for field station closures, although natural disasters and lack of community support also may play roles. Winter 2012 was the first time since 2005 during which the Polar Environment Atmospheric Research Laboratory (PEARL) in Eureka, Nunavut—the northernmost permanent, nonmaterial research facility in the world—did not take any scientific measurements. PEARL ceased year-round operations in April 2013 when it lost its federal funding despite the Canadian government's assertion that Arctic research has high priority for Canada. PEARL now operates part-time on a donation basis (Globe and Mail, 2013). In 2011, the University of Manitoba closed the 45-year-old Delta Marsh Field Station because of severe damage caused by spring flooding. The cost of repairs, the reclassification of the land as floodplain, and the subsequent inability to secure new flood insurance led to the University's decision to not reopen the station (CBC News 2011). Lack of financial or community support has caused other field stations (e.g., the San Blas Field Station in Panama; the National Wildlife Research Center in Kingsville, Texas; and the Meanook Biological Research Station in Manitoba, Canada) to close permanently (Alper 1998, Annand 2014, USDA 2014).

Conclusions and Report Roadmap

The research conducted at field stations is rich in diversity and depth, and is respected for moving science forward in fundamental ways that have changed our view of nature and advanced ecological theory. Field stations constitute an important part of a nation's research infrastructure, one that enables scientists to better understand the world's complex natural history and socioenvironmental systems, and to better measure the rapid environmental changes that are stressing natural and social systems.

Sustained support for field stations allows continued access to a diverse array of ecosystems in which scientists can conduct reasonably protected long-term studies and manipulative experiments that are crucial if we are to understand the environmental, ecological, and evolutionary causes of observed changes on large scales of time and space. If field stations are to thrive in the 21st century and beyond, they will need to become more flexible, better able to adapt to changing research technologies, to changing economies, and to the changing environment in which we all are embedded. The following chapters outline a course of action to make that possible.

Chapter 2 describes strategies for increasing the value, relevance, and sustainability of field stations while enhancing their ability to adapt to changing environments, research technologies, and economic conditions. Chapter 3 presents opportunities to support these strategies through networking. Chapter 4 focuses on the challenges and opportunities to build and maintain infrastructure. Chapter 5 argues for visionary leadership and financially sustainable business models for field stations. Chapter 6 addresses the need for field stations to develop and document their impact. This requires collecting the necessary data and making them accessible to allow for trend and impact analyses across the community of field stations.

2
Enhancing Science, Education, and Public Engagement

You never change something by fighting the existing reality. To change something, build a new model that makes the existing model obsolete.
—R. Buckminster Fuller, 1975

The rich foundation of basic natural-history research conducted at field stations demonstrates their capacity to foster innovative and synergistic science. To survive in the future, field stations should also address important societal issues. Many reports have identified emerging environmental challenges that need to be addressed, and some have outlined strategic scientific plans for addressing them (NRC 2001, 2009b). A report of the National Association of Marine Laboratories and the Organization of Biological Field Stations (Billick et al. 2013) also outlined emerging environmental trends and developed goals and actions to maximize the use of field stations and marine laboratories to address those trends (Box 2-1). As described in the NAML-OBFS report, collaborative research is a hallmark of field station work and will be required in increasing measure to address the complex scientific questions facing society.

Field stations generally are strongly committed to the conduct of science, education, and public outreach in their mission statements. Some are informally connected through professional societies and regional research networks, but they do not have a unifying management structure. The ability to use field stations effectively to address important environmental and societal issues will require new and enhanced models of collaboration and networking and strategic business plans that are integrated with and as robust as their strategic scientific research plans. This chapter focuses on new models of collaboration that can help to fill knowledge gaps in science and engineering, support innovative policy decisions required in the face of global environmental change, and encourage and empower the public to engage in and support scientific endeavors.

Promoting Convergence

New approaches to collaboration are needed to address knowledge gaps in biology, earth-systems science, environmental science, and engineering. Field stations offer unparalleled opportunities to address questions that fall between or span the domains of traditional scientific disciplines and academic departments.

Field stations have a long history of bringing multiple disciplines together to address scientific and societal challenges. Common practice at many field stations

BOX 2-1
NAML-OBFS Report, Field Stations and Marine Laboratories of the Future: A Strategic Vision

In 2013, the National Association of Marine Laboratories (NAML) and the Organization of Biological Field Stations (OBFS) jointly published a strategic vision report to guide, improve, and demonstrate the scientific and educational value of field stations and marine laboratories (FSMLs) to broader society. To develop their report, NAML and OBFS sought guidance from the field station community through a public workshop (NAML-OBFS 2013a) and a broad-scale survey (NAML-OBFS 2013b) that took stock of the perceived strengths, limitations, needs, and ideas for improvement of field stations around the world. The OBFS/NAML report recommended the following four strategic goals:

Goal 1 Increase the value to society of the science done at FSMLs, as well as the public understanding of that value.

Goal 2 Increase the scientific value of FSMLs by increasing the flow of information, both between [field stations] and scientists and among FSMLs themselves:

Objective A Develop a more comprehensive network of FSMLs.
Objective B Increase the ability of scientists to take advantage of FSMLs.

Goal 3 Enhance the synergies between research and education.

Goal 4 Promote the flow of scientific information for environmental stewardship by ensuring appropriate access by scientists and students to terrestrial, aquatic, and marine systems.

Goal 5 Increase the operational effectiveness of FSMLs:

Objective A Enhance the effectiveness of individuals working at FSMLs.
Objective B Maintain and improve critical infrastructure.

SOURCE: Billick et al. 2013 (pp. 36-40).

---

is the merger of different areas of expertise to address knowledge gaps by fostering interchange among communities that include natural and social scientists, educators, private-sector professionals, and society at large. A promising trend in the scientific community that embraces collaborative and multidisciplinary methods of inquiry to address daunting and urgent challenges has been given a name: *convergence* (Box 2-2).

Convergence of the life, physical, computational, and mathematical sciences is resulting in transformational paradigms for scientific and technological advances (Sharp et al. 2011, American Academy of Arts & Sciences 2013, Roco et al. 2013, NRC 2014a). As universities, industries, and funding organizations grapple with how to facilitate scientific convergence, field stations are positioned to contribute to the movement. Convergence to address societal challenges is an important pathway in science research that field stations can use as they strive to meet Goal 1 of the NAML-OBFS Strategic Plan. A National Research Council report (NRC 2014a) describes in detail barriers to, strategies that facilitate, and characteristics of successful programs for convergence. Some of the strategies that institutions have

BOX 2-2
Definition of Convergence

Convergence is an approach to problem solving that cuts across disciplinary boundaries. It integrates knowledge, tools, and ways of thinking from life and health sciences, physical, mathematical, and computational sciences, engineering disciplines, and beyond to form a comprehensive synthetic framework for tackling scientific and societal challenges that exist at the interfaces of multiple fields. By merging these diverse areas of expertise in a network of partnerships, convergence stimulates innovation from basic science discovery to translational application. It provides fertile ground for new collaborations that engage stakeholders and partners not only from academia, but also from national laboratories, industry, clinical settings, and funding bodies.

SOURCE: NRC 2014a (p. 1).

taken to promote convergence that may be familiar practices to many in the field station community include the following:

- organizing research programs around common themes or scientific challenges
- fostering opportunities for researchers to interact
- changing existing faculty structures and reward systems
- working with and across existing departments
- designing facilities and workspaces for convergent research
- designing education and training programs that foster convergence.

Essential elements of successful convergence programs include people, organizational structure, culture, and research ecosystems (Box 2-3). Notable among the strengths of a field station community are the people involved in station programs and activities and the station culture. People of all ages come together at field stations, and this fosters a thriving "station culture," which in turn promotes a collaborative environment that can lead to serendipitous scientific discovery (Michener et al. 2009). The space and time to nurture cross-generational and cross-disciplinary relationships is a valuable component of convergence that is afforded at field stations where undergraduates, graduate students, postdoctoral students, faculty, and others may interact for weeks or months, year after year. This advantage is not lost at field stations close to or embedded in population centers. Field station users often interact with their surrounding communities. Hence, local governments and community members are more likely to play a role in identifying local scientific challenges that could serve as focus areas for field station research. Research focused on local issues may also encourage citizen participation in science. Local knowledge about wild species, landscapes, and human culture can be an important contributor to scientific research, and the use of local knowledge in ecological research is on the rise (Brooke and McLachlan 2008).

Field stations have the capability to nurture the formation of transdisciplinary research groups that address cross-cutting scientific questions and urgent societal concerns, and many have done so for decades. The formal codification of this approach into convergence opens the opportunity for station leadership to highlight and to strengthen these activities at field stations.

Managing and conserving ecosystems require incorporation of perspectives of disciplines beyond life and physical sciences and engineering—disciplines such as economics, demography, and the humanities (Ewel 2001). Despite the positive trend toward convergence, the social sciences, arts, and humanities—which have much to contribute—often remain underrepresented. Field stations provide a setting for natural and physical scientists, social scientists, humanists, and artists to come together and collaborate. The Ecological Reflections project—which brings environmental science, arts, and humanities together—has been particularly effective in bringing artists to field stations.[10] The artists and humanists explore the cultural and moral meanings of nature and place in settings as diverse as the old-growth conifer forests of the Oregon Cascades, the north temperate lakes of Wisconsin, the Minnesota prairie, and the saguaro desert in Arizona. Artists-in-residence programs enhance the education and research activities of many field stations and lead to more innovative science and a greater understanding of the sociocultural consequences of environmental change (Ewel 2001; Sorlin 2012). Thus, some field stations have demonstrated the capability to encourage convergence not only of the life and physical sciences, but also of the social sciences and the arts and humanities.

The organization and ecosystem of partnerships (see Box 2-3) are areas where many field stations need strengthening—for their long-term viability as well as to support convergence.

As field stations become more networked and as distributed partners coordinate their efforts, the ability for scientific staff at field stations to address societal concerns (such as species invasions, fire behavior, water storage and cycling, and carbon sequestration) at a variety of scales from regional to national to global can be enhanced by a greater emphasis on convergent research. Networks of field stations have the potential to become Earth-scale test beds for developing and testing new monitoring technologies and sustainability practices (NRC 2009b, Roco et al. 2013).

Positioning field stations to address societal challenges with a broad-scale, convergence-driven approach is not a simple undertaking. For many stations, such a shift will require financial resources, enhanced infrastructure, and networking with other field stations and other kinds of institutions. Successful strategies will require careful consideration of how to leverage existing resources and infrastructure efficiently and effectively in addition to building new ones. Chapters 3 and 5 address the networking and financial needs, respectively, in more detail.

Expanding and Diversifying Discovery-Based Learning

The unity of all knowledge, "the linking of facts and fact-based theory across disciplines to create a common groundwork of explanation," was captured by E.O.

---

[10]http://www.ecologicalrefelections.com

BOX 2-3
Essential Cultural and Structural Elements in Successful Convergence Ecosystems

People: A commitment to supporting convergence from all levels of leadership is key, as is the involvement of students, faculty members and staff, department chairs, and deans.

Organization: Inclusive governance systems, a goal-oriented vision, effective program management, stable support for core facilities, and flexible or catalytic funding sources are all critical to organizations seeking to build a sustainable convergence ecosystem.

Culture: The culture needed to support convergence, as with other types of collaborative research, is one that is inclusive, supports mutual respect across disciplines, encourages opportunities to share knowledge, and fosters scientists' ability to be conversant across disciplines.

Ecosystem: The overall ecosystem of convergence involves dynamic interactions with multiple partners within and across institutions, and thus requires strategies to address the technical and logical partnership agreements required.

SOURCE: NRC 2014a (pp. 8-9).

Wilson with the term "consilience" (Wilson 1998). Consilience is a natural complement to convergence. Student, science professional, and citizen exposure to unifying theories across disciplines is needed in order to tackle and solve pressing scientific and societal challenges.

Social science research demonstrates that active learning,[11] which includes discovery-based learning, particularly through early research experiences, advances student persistence and success in science, technology, engineering, and mathematics (STEM) disciplines, especially in women and other underrepresented groups (Nagda et al. 1998, PCAST 2012, Graham et al. 2013). Active learning enhances students' ability to solve problems, an essential skill that is needed to address pressing societal environmental challenges (Hake 1998, PCAST 2012). Student research and other active learning experiences also improve grades, increase student self-identification as scientists or engineers, reduce the time to graduation, and increase interest in postgraduate education (Seymour et al. 2004, Lopatto 2007, Santer 2010, Dirks 2011). Field stations already play a critical role in exposing students—from elementary school to high school to college—to the natural environment and getting them excited about science. Empirically based approaches to education will be important as field stations strive to meet Goal 3 of the NAML-OBFS Strategic Plan, to enhance the synergies between research and education. Educational programs at field stations would be enhanced by embracing and implementing the findings and guidelines on active learning from the education-research community.

---

[11] Active learning is a student-centered approach to instruction that requires students to engage in meaningful learning activities (Dirks 2011). Discovery-based learning is one of a range of "active learning" approaches (Michael 2006).

Research at field stations, being hands-on and embedded in the environment, naturally lends itself to discovery-based research experiences for students. As universities focus on increasing student recruitment into and retention in STEM disciplines, the time is ripe for field stations to create educational research programs that benefit not only students in earth science, environmental science, and ecology, but students in other STEM disciplines as well. By expanding opportunities at field stations for independent and collaborative research projects, and perhaps by moving some undergraduate laboratory courses from campuses to field stations and adding a field component to them, universities might increase students' interest in pursuing STEM disciplines.

Discovery-based learning at field stations should not be limited to STEM majors. Social science and humanities programs can also attract students to field stations. Science can be made both real and relevant to social science students by allowing them to study at field stations for a semester-long or summer program. In addition, bringing together STEM, social science, and humanities majors at field stations could nurture rich educational experiences for students in all groups.

A good example of the integration of the arts and sciences at field stations can be found at the Mountain Lake Biological Station (MLBS) in Virginia. The MLBS ArtLab program brings artists and scientists together to "share viewpoints, observations, philosophies, and perspectives in their common quest to observe and understand nature and biology" (MLBS 2014). In 2013, the MLBS hosted its first artist-in-residence, recognizing that the field station's setting would serve as a great inspiration for those working in creative arts. Another example is the Logan Science Journalism Program[12] at the Marine Biological Laboratory in Woods Hole, Massachusetts. The Science Journalism program is an opportunity for communication professionals to learn about and engage in basic research, thus improving their understanding of the process of science. Field stations should expand such opportunities for social science and humanities majors and professionals, to enable them to learn and create in settings where scientists have been working for decades.

As they build and expand discovery-based learning programs in STEM and social science disciplines, field stations should partner with the education research community. They can be platforms for research on how people learn. An example is the partnership between Oregon Sea Grant's Free-Choice Learning Laboratory and the Hatfield Marine Science Center (OSU 2014). Field stations should look for opportunities to develop similar research partnerships.

Actively Engaging the Public in Science

Field stations carry out a wide range of public engagement activities to improve public access to and understanding of science. Public understanding and participation in science are important to increase human connectedness to the natural world and empower citizen decision making and involvement in public

[12] http://www.mbl.edu/sjp

policy (Brossard and Lewenstein 2010, Fischoff 2012, Nadkarni and Stasch 2012). Most scientists and research institutions "communicate" about science through peer-reviewed journal publications (Harley 2013), which reach primarily scientific audiences. The outreach activities of field stations break this mold, and the committee applauds such efforts.

Over the last 30 years a small, but robust research base has been built on effective science communication and informal education (Brossard and Lewenstein 2010; Fischoff 2012; NRC 2009a, 2014b). Infusing principles that stem from science communication and informal education research into engagement activities at field stations may help to enhance the relevance, effectiveness, and thus the long-term sustainability of their outreach programs. Use of empirically based approaches will also propel field stations toward achieving Goal 1 of the NAML-OBFS Strategic Plan, to increase public understanding of the value of field station research to society. Science-based approaches to outreach activities can also create opportunities to train students and early-career scientists in public engagement. Four important tasks for developing effective science communication activities are: (1) identify the science relevant to decision making, (2) determine what people already know, (3) design communications to fill the critical gaps, and (4) evaluate their adequacy and repeat as necessary (Fischoff 2012). The National Research Council report, *Learning Science in Informal Environments* (NRC 2009a), outlines six interrelated "strands of science learning" that form a framework for "science-specific capabilities supported by informal environments" and "serve as a conceptual tool for organizing and assessing science learning (Box 2-4). Central to both empirically based science communication and informal education is to first listen to and understand what people value and to evaluate the effectiveness of engagement activities given what is known about people's values. The Center for the Advancement of Informal Science Education[13] is an informal science education resource for many institutions involved in public engagement activities, including field stations. Many of the principles of developing effective public participation in science activities are also relevant to formal education.

Citizen science is one powerful channel through which field stations can engage and empower the science-interested public and advance science. The democratization of science, enabled by the general public's increasing access to information and tools that were once the exclusive domain of experts and specialists has enabled citizens to become increasingly involved in the collection and analysis of biological and environmental data. Those data, in turn, are increasingly being transformed into scientific information and understanding, critical at a time when public understanding of science concepts and processes is disturbingly low (Miller 2007). Citizen science facilitated or hosted by field stations constitutes a potential win-win scenario: an engaged public may be better at understanding, appreciating, and supporting how scientific knowledge is acquired

[13] Center for the Advancement of Informal Science Education (CAISE) serves as a resource for strengthening and advancing the field of professional informal science education (http://informalscience.org/). CAISE works in collaboration with the NSF Advancing Informal STEM Learning Program and the Association of Science-Technology Centers.

BOX 2-4
Six Strands of Science Learning

Learners in informal environments:

Strand 1: Experience excitement, interest, and motivation to learn about phenomena in the natural and physical world.

Strand 2: Come to generate, understand, and remember, and use concepts, explanations, arguments, models, and facts related to science.

Strand 3: Manipulate, test, explore, predict, question, observe, and make sense of the natural and physical world.

Strand 4: Reflect on science as a way of knowing; on processes, concepts, and institutions of science; and on their own process of learning about phenomena.

Strand 5: Participate in scientific activities and learning practices with others, using scientific language and tools.

Strand 6: Think about themselves as science learners and develop an identity as someone who knows about, uses, and sometimes contributes to science.

SOURCE: NRC 2009a (p. 4).

at field stations and applied, and scientists' research may be enhanced by the intellectual and data input from an engaged public.

Although some field stations are actively engaged in citizen science initiatives, there are many ways in which field stations could expand and enhance these initiatives. There is a large potential range of approaches to promoting citizen science—from smaller, place-based programs that investigate relevant questions on site, to large-scale, existing programs for which a field station may facilitate one of many nodes of input and can include inputs from an expanding community of citizen scientists. Field stations seeking to add a citizen science component to their monitoring programs have more tools and resources at their disposal than ever before. Web applications, social networks, and digital games are some of the new digital tools that are facilitating citizen science projects (Bowser and Shanley 2013). In addition, new developments in information science—including data informatics, graphical user interfaces, and geographic information system applications—can now be used on smartphones, tablets, and personal computers. For example, eBird is a large-scale citizen science program that engages thousands of volunteers in documenting millions of bird observations (over 3 million in 1 month alone in 2012). eBird encourages users to participate by providing Internet tools to maintain their personal bird records and to visualize data with interactive maps, graphs, and bar charts, which allow rapid access to the records in the field. The embedding of environmental sensors in smartphone technology and wearable accessories are additional technological advances in a rapidly growing commercial enterprise that has parallel scientific applications. The distributed networks of mobile sensors combine citizen science with health monitoring and systems analysis and ideally could be tested by the field station community (Zhang et al. 2011).

Citizen science is an increasingly important component of environmental monitoring and public engagement with the scientific community. Collaboration and connection with other government initiatives—such as America's Great Outdoors, [14] which engages volunteers and citizens, especially youth—have been noted as having substantial benefits both for environmental monitoring and conservation and for society as a whole. LiMPETS (Long-term Monitoring Program and Experiential Training for Students) is an environmental monitoring and education program for students, educators, and volunteer groups that was developed to monitor the ocean and coastal ecosystems of California's National Marine Sanctuaries to increase awareness and stewardship of these important areas. About 4,000 teachers and students along the coast of California are involved in the collection of data on rocky intertidal beaches and sandy beaches as part of the LiMPETS network.[15] Citizen science initiatives have the ability to enable coordinated networks of volunteers to collect useful data that can inform our understanding of the state of ecosystems.

Field stations are places where citizen science can be encouraged, where cooperation in the collection and understanding of data can be collectively transformed into an understanding of the environment and expressed in ways that are relevant and important to public audiences. Citizens do not replace scientists, but can contribute to the vast array of environmental data and information that are needed to study and understand our changing world, from species identification, to water-quality and air-quality monitoring, to building networks for early detection of environmental change. In addition, field stations could collaborate to develop coordinated networks of citizen science monitoring programs at national and international levels. The *New Visions in Citizen Science* report (Bowser and Shanley 2013) outlines evidence of impact and approaches to address challenges for 17 case studies of citizen science projects that may be instructive to field stations seeking to build such programs. The Citizen Science Association (CSA)[16] is another resource that field stations might consider to foster the development of their citizen science programs and incorporate best practices. Because the CSA was just formed in 2014, leaders in the field station community have an opportunity to be inaugural members and to help the CSA define its scope and direction.

Overcoming Barriers

Programs to implement convergence, to develop interdisciplinary-based education opportunities, and to enhance public outreach do not come without challenges. Tenure and promotion criteria can be impediments to young

---

[14] http://www.doi.gov/americasgreatoutdoors/index.cfm
[15] http://limpetsmonitoring.org
[16] http://citizenscienceassociation.org

researchers interested in public outreach activities or collaborative research programs such as convergence. Teaching space and equipment and transportation are important elements to consider for education programs. Data quality, trained personnel, and liability for the safety of volunteers are common challenges in citizen science programs. The need to restrict access to sensitive ecosystems can be an impediment to both formal education and public outreach activities. To move forward, each field station should consider how to tailor programs given its facility, location, personnel, and other available resources. Field stations will also need to consider whether changes are needed in their organizational or cultural infrastructures. Networking, cyberinfrastructure, and business planning, as discussed in the next chapters, will be important elements for overcoming a variety of barriers.

Conclusions

Field observations have played and will continue to play an important role in the physical, natural, and social sciences. Field stations collectively constitute a critical global asset with the potential to facilitate a unique merger of natural capital, intellectual capital, social fabric, and infrastructure that lead to important scientific research required to understand our rapidly changing natural world. Greater emphasis needs to be placed on cross-disciplinary research, including research in the geosciences, the social sciences, the humanities, and the arts.

Sustained infrastructure support for field stations allows access to historical data, (longitudinal data that have not been archived in databases) and to long-term studies and manipulative experiments that are unique to specific ecosystems and that enable us to understand the driving forces behind environmental change. A greater emphasis on convergent research that includes the geosciences, social sciences, humanities, and arts will enhance scientists' use of historical datasets to address global challenges. The recognition of the importance of this set of activities in what is now called convergence is an identified strength of many field stations.

> ***Recommendation***: Field station leaders should identify and support the development of scientific and educational assets that harness their station's unique qualities to address local, regional, national, and global challenges by bringing together scientists in a number of disciplines, including the social sciences, through what is now called convergence.

Recruiting students into STEM fields has been identified in many nations as having high priority, given the importance of these fields for innovation and economic growth. Field stations are venues for discovery-based learning and offer rich opportunities for other types of active learning that have been shown to promote diversity and persistence in STEM education. Integration of research into formal and informal education and public engagement in science activities provide engaging learning opportunities for people of all ages and backgrounds.

> ***Recommendation***: Universities and other host institutions should expand opportunities at field stations for independent and collaborative research and active learning activities to increase interest and persistence in STEM fields.

Public understanding and participation in science is important to increase human connectedness to the natural world and to empower citizen decision making and involvement in public policy. Field stations support a wide range of public outreach and engagement programs—public lectures and workshops, science cafes, field trips, and nature walks, among other informal education opportunities—to enhance public understanding of science. The committee applauds these public outreach efforts because they break the mold of traditional science communication with programs that more actively involve members of the public in science. However, field station outreach programs are often disconnected from empirically based approaches to develop, evaluate, and document the effectiveness of their science communication and informal education activities.

> ***Recommendation:*** Field stations should continue to explore a wide range of approaches to engage the public in science, and to select and tailor their activities in a manner that best leverages a field station's infrastructure, location, personnel, and other available resources. Empirically based approaches in science communication and informal learning should be used to guide the development and assessment of engagement activities to promote public understanding of science effectively.

Citizen science is an emerging channel through which field stations can advance science and empower people interested in science by engaging them actively in data collection and research, particularly in science issues that affect their communities. There is a broad spectrum of citizen science initiatives, from simple observational programs to coordinated, training-intensive environmental monitoring programs. Citizen science initiatives empower people to learn about science and the ecosystem dynamics of the natural communities in which their field stations are embedded. Citizen science initiatives also can enable coordinated networks of volunteers to collect data that can inform our understanding of how human activities may be altering ecosystems. Much of citizen science is facilitated through advances in Web-based technologies that allow citizens to collect and analyze data through accessible platforms, such as smartphones, tablets, and personal computers. A few field stations have developed sustained outreach programs that include citizen science, but citizen science initiatives are not yet widespread outreach activities among field stations.

> ***Recommendation***: Field stations should collaborate in, connect with, and formalize citizen science programs by using the latest technologies and

networking initiatives throughout the American and global system of stations and thus offer a coordinated infrastructure for interested members of the public to engage in, learn about, and contribute to science.

3
Networking Field Stations for Discovery and Innovation

> *Self-organizing networks that span the globe are the most notable feature of science today. These networks constitute an invisible college of researchers who collaborate not because they are told to but because they want to, who work together not because they share a laboratory or even a discipline but because they can offer each other complementary insight, knowledge, or skills.*
>
> —Caroline Wagner, 2008

Field stations can enhance their contributions to research, education, and outreach through their research initiatives, linking their data-sharing portals, and partnering with similar institutions. The global distribution of field stations suggests enormous potential for them to become core components of an Earth-scale environmental neural network that contributes to monitoring, preparing for, adapting to, and training future generations to address environmental change. Such a neural network will require connections that go beyond membership affiliations with professional societies, although professional societies can play supporting organizational roles.

What Is a Field Station Network?

Existing field station networks and collaborative efforts range from informal associations among scientists, such as the Global Lake Ecological Observatory Network[17] and the Nutrient Network,[18] to more formal consortia that collect data on a variety of ecological processes, such as the Long-Term Ecological Research Network[19] (LTER) and the National Ecological Observatory Network[20] (NEON). The National Estuarine Research Reserve, [21]supported by partnership between the National Oceanic and Atmospheric Administration coastal U.S. states and, is another example of an existing network. Some field stations form networks around common research efforts or to share resources, including research protocols, research equipment, or educational curricula. Examples of such networks are the University of California's Natural Reserve

[17]http://www.gleon.org
[18]http://www.nutnet.umn.edu
[19]http://www.lternet.edu
[20]http://www.neoninc.org
[21] http://www.nerrs.noaa.gov

BOX 3-1
What Is a Network?

A network is a set of nodes (people, places, or institutions) connected via ties, such as exchange of information, resources, or activities (van Alstyne 1997, Borgatti and Foster 2003). Many different forms of networking can occur, from simply sharing information and data to far more formal, regionalized to international groups of field stations. Field stations tend to build innovative, unique networks of scientists from varying backgrounds who would otherwise rarely interact, except for the social and scientific exchange that goes on at field stations (Michener et al. 2009).

System,[22] the Partnership for Interdisciplinary Studies of Coastal Oceans (PISCO[23]), and the Consortium of Universities for the Advancement of Hydrologic Science, Inc. (CUAHSI[24]).

Networks of researchers result in communities of researchers, often from around the world, linked by virtual ties, whose desire to collaborate is fueled by shared interests to advance science, not institutional mandates. Such networks form, mutate, dissolve, and reform, bringing together scientists of diverse backgrounds who offer each other the benefits of their insights, knowledge, and skills (Wagner 2008). The LTER network is an example of a relatively stable research network with 27 sites, some of which are field stations that benefit from intranetwork coordination and comparison. The coordination and information sharing are facilitated greatly by a central network office, data management system, and scientist meetings at regular intervals.

Despite those examples, many field stations still operate independently and in isolation from one another. Field stations would benefit substantially from networking with each other and with national parks, wildlife refuges, estuarine research reserves, and other research centers. This would provide novel opportunities to enhance research capacity and financial efficiency while sparking innovation and opening new arenas of scientific inquiry, education, and outreach (Box 3-1).

The most effective networks are self-defining and self-organizing, where a reciprocal exchange of goods or services takes place (Wagner 2008). In this report, the committee considers a range of networks from informal to formal including the following:

- scientists sharing ideas, data, and best practices
- scientists collaborating on research efforts across multiple sites
- institutions sharing organizational efforts and resources
- collaborations and partnerships between field stations, public agencies, nongovernmental organizations, and industry

---

[22]http://nrs.ucop.edu
[23] http://www.piscoweb.org
[24] http://www.cuahsi.org

Principles, Benefits, and Challenges

The desirability of connecting sites of long-term ecological research has been recognized broadly (Billick et al 2013). Networking offers the benefit of connecting place-based knowledge over larger geographic scales, which can be the impetus for establishing an increasing number of networks (Schimel et al. 2011). The need for and benefits of sharing and comparing discoveries and data among sites will continue to grow in the 21st century. For example, an important scientific question for the 21st century is: What changes must human societies make to adapt to rapid and unpredictable environmental changes while maintaining resilient systems? Adaptation, threshold, and resilience research have high national priorities. Addressing such applied-research questions would require a network of practitioners and scientists that can access long-term datasets, place-based information and knowledge, and resource management expertise (NRC 2010a). Field stations can contribute fundamentally to building such a network; much of the needed investment has already been made.

There is increasing evidence that well-designed networks spark innovation, spread ideas, lead to smarter decisions and greater efficiency, and even have measurable effects on local gross domestic product and the number of new patents (Pentland 2014). While networks of scientists and field stations may form organically, network theory and analyses suggest three principles that can enhance their value. First, if incentives for forming a network are offered, the results can be dramatic. The second principle is that networks need both local clusters and long-distance connections for leapfrogging ideas. In the field station context, this implies local clusters of field stations within regions with links to a broader national network. The third principle is that diversity of network nodes enhances innovation and scientific breakthroughs (Pentland 2014). Those principles support arguments against closing networks because of restrictive data requirements.

Networking offers other significant advantages: (1) networks can capture sufficient intellectual capital—a range of scientific and other knowledge—to tackle cutting-edge research questions and seize opportunities that no single field station could do alone; (2) networks can attract the intellectual capital that enhances creativity and innovation while creating opportunities for multidisciplinary collaboration and convergence; (3) networks can facilitate resource pooling to make investments in large infrastructure more efficient, such as data and information management (including new tools for mining, analyzing, and visualizing data), cyberinfrastructure, and analytical laboratory equipment; and (4) networks can facilitate research coordination, reducing redundancies in research projects. Field stations and the resulting science would benefit greatly from coordinated and standardized data management protocols and data portals.

Field stations could become nodes for development of regional clusters that include other research centers or sites for particular environmental challenges and research. Gap analyses that assess whether particular ecosystems are well represented within a network could help guide selection of sites and partners. The common foundation provided by NEON sites creates data hubs that could be expanded into a more extensive and comprehensive environmental sensing network (this more extensive vision is not currently part of NEON's mandate, and incorporating additional nodes would come at a cost). For example, LTER sites—many of which are at field stations—could become additional nodes, thus adding a historical context and long-term datasets to extend NEON records in time and space. A robust, comprehensive environmental sensing network would require strategies to become more inclusive by supporting the web presence and data storage for field stations or other approaches that would accomplish regional data integration across NEON, LTER, field stations, and other research centers—particularly aquatic research that explores connectivity between terrestrial, coastal, and ocean ecosystems. Longitudinal data and natural-history observations collected at field stations are complementary to and could help explain and interpret the data collected at NEON. A more comprehensive network strategy that includes field stations would add richness and depth to large-scale environmental observation networks.

Global-change research requires infrastructure that includes

- long-term ecological and environmental datasets that allow detection of changes on a variety of spatial and temporal scales,
- a broad biogeographic sampling with replication of ecosystem types and disturbance gradients with which to track change and resilience (or lack thereof),
- a legacy of manipulative ecological experiments that can be repeated to assess how fundamental ecosystem properties are altered by controlled perturbations and how they recover once the disturbances are removed and;
- a flexible platform for multidisciplinary collaboration among scientists.

Networking is valuable not only for science but for education and outreach. Individual field stations and their K-12 audiences benefit from sharing K-12 curricula. Several networks have already been established for the purpose of sharing educational and outreach knowledge (e.g., Coastal America's Coastal Ecosystem Learning Center Network (CELC), Centers for Ocean Sciences Education Excellence (COSEE), and Communicating Ocean Sciences—Reflecting on Practice) Indeed, once developed, innovative curricula and curricular modules and research projects that focus on societal priorities can be readily adapted to lectures and public interpretive programs among all the field stations in a regional network. By including other ocean- and land-management organizations (nongovernment organizations and local, state, and national government agencies), field stations could effectively develop and make available educational information on invasive species, wildland fire and fuel management, floods and droughts, and other perhaps region-specific research subjects that are not necessarily parts of the research portfolio of an individual field station, but that have broad public interest and importance.

Data from student projects and citizen science projects could be shared among stations in a network and with other research organizations that are engaged in similar work. Linking would make it possible for field stations to distribute the costs and management of collaborative projects, and it would be much easier for all types of students and citizen scientists to learn and conduct research at any of the member field stations in a region or around the world. The concept is somewhat akin to the America the Beautiful—The National Parks and Federal Recreational Lands Pass,[25] which allows entry into all national parks with just one card. For those field stations that are able and willing to participate, criteria could be established for field station access by K-12 students, high school faculty, citizen scientists, and other members of the public.

Although many field stations are not yet organized into science-based networks, respondents to a survey of the National Association of Marine Laboratories and the Organization of Biological Field Stations (OBFS) ranked collaboration and networking as the two greatest public benefits that field stations could provide (NAML-OBFS 2013b). Over 40 percent of the respondents reported that their field station is involved in some type of public outreach or engagement activity—spanning traditional public outreach, consultation with industry, community mediation, and environmental policy mediation and advising. Comments indicate that those field stations are highly engaged in their communities and responsive to their communities' needs and opportunities. Field stations often pursue formation of local and informal networks.

Networking can also help in raising funds from nontraditional sources. For example, foundations often have specific funding priorities for projects that either are topically or regionally based. Available grant funds vary from a few thousand dollars to millions of dollars. The Leona M. and Harry B. Helmsley Charitable Trust[26] has expressed interest in funding large projects at field stations that will have national impacts. Other foundations might be convinced that a well-organized network of field stations dedicated to supporting research and training for future scientists is worthy of support.

Research and watershed and near-shore resource management could be improved by networks of field stations. Land management can be expensive and for many objectives can be effective only when all the stakeholder groups surrounding a field station participate. Exotic species, invasive plants, and nonnative fish, for example, can be effectively managed only when there is strong involvement by representatives of all groups in the local watershed. The same is true for wildland and prescribed fire management. Field stations can provide data on wildland fuel conditions, local weather, and historical fire regimes, and trained staff and training sites (such as the Archbold Biological Station[27] and the Tall Timbers Research Station.[28]). Furthermore, shared best management practices, staff, and equipment can all improve the outcomes. For example, obtaining meaningful data on watersheds often requires many sampling points using a

[25] https://store.usgs.gov/pass/index.html
[26] http://helmsleytrust.org
[27] http://www.archbold-station.org
[28] http://www.talltimbers.org

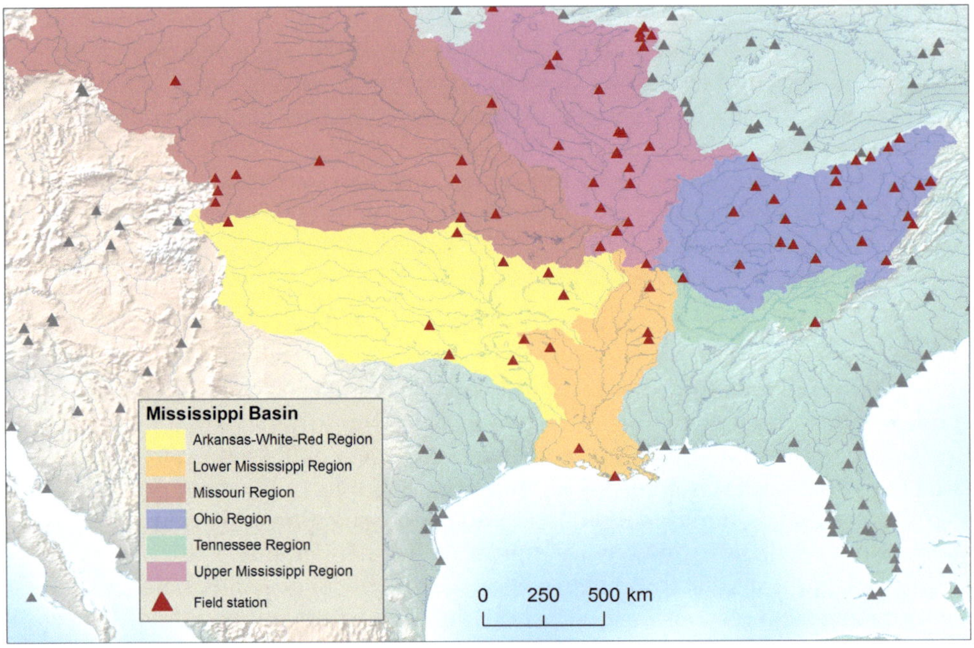

FIGURE 3-1 Field stations within the Mississippi River Basin. The Mississippi River watershed spans many states and river systems, as indicated by the color blocks. Field stations located within the river basin are marked in red.

common methodology and data processing. Many field stations in the United States are strategically located within major watersheds. See, for example, the map of field station locations in the Mississippi River watershed shown in Figure 3-1. A network of these field stations could collaborate in designing and conducting a monitoring and research program to evaluate nutrient loading and eutrophication of the Gulf of Mexico and its dead zone. This network could serve as a model for assessing the influences of land use on coastal waters and the connectivity of freshwater and marine ecosystems.

For nearshore marine sites, the combined efforts of numerous marine laboratories produced useful research to guide fisheries and intertidal management.[29] Networking has also improved the brick and mortar infrastructure of some field stations.

Sustainable and cost-effective construction at some field stations has benefited from the OBFS members who share their data on green buildings and energy efficiency.[30] Networks also can help to distribute demand for access to provide protection of particularly sensitive natural areas.

[29] http://www.piscoweb.org/topics/marine-resources

[30] http://www.obfs.org/assets/docs/obfs_sustainprinciples.pdf

Despite the benefits of networking, the challenges should not be underestimated (Wagner 2008). Introduction and implementation of new organizational structures and activities, such as networks, require effort and leadership to overcome hurdles such as resistance to change, vulnerability to participant withdrawal, and costs of maintaining a network. Most field stations operate at near capacity, and taking on a partnership in a network may require additional human or financial resources. The level of cyberinfrastructure at field stations, including cellular communication, adequate data transmission capabilities, computer resources, and video conferencing, varies significantly and is a barrier that needs to be addressed. In addition to having the infrastructure needed to enable connectivity and participation, care must be taken in designing and nurturing networks to ensure that field stations do not lose their individuality and core missions.

Building and Establishing Networks and Partnerships

The most successful and sustainable networks tend to be those that are self-organizing and self-defining (Wagner 2008). Those that are introduced or implemented by leaders using a top-down approach often meet with resistance, are difficult to start, and are even more difficult to sustain. Network theory and analyses provide three strategies that can ameliorate those problems while enhancing the acceptance and overall value of networking: (1) offer incentives to institutions to form networks, (2) form networks with both local clusters and long-range connections for leapfrogging ideas, and (3) include diversity among network nodes to enhance innovation and scientific breakthroughs (Pentland 2014).

In developing a network, each participating institution should identify the special value that it brings to the network and derives from it (Wagner 2008) (Box 3-2). Reciprocity among network members is essential for sustained success. Sometimes field stations affiliated with federal or state governments can share activities by simply providing access to protected lands.

Small informal networks often arise and persist for as long as their original defining needs exist. Such networks should be encouraged, advertised, and supported, perhaps through a "small-network" grant program. With the emergence of user-friendly online tools for networking and data sharing, the cost and effort of beginning the process of networking have been lowered. In contrast, larger, top-down networks that are developed by institutions may require substantial management effort and financial support and need to be developed carefully with particular goals and incentives in mind.

Financial incentives might be needed to prescribe network formation.. For example, less than a decade after its formation, the European Union (EU) managed to create cross-boundary, collective research endeavors called European Research Areas.[31] Financial incentives were used to help remove the narrow, within-

[31] http://ec.europa.eu/research/era/index_en.htm

BOX 3-2
The Utah Field Station Network, a Regional Network to Enhance Research and Education

Researchers and students at Lytle Ranch Preserve in Utah. Photo Credit: Bryan Adams, Brigham Young University.

The Utah consortium links field stations that are administered by universities and state and federal agencies. The consortium is "dedicated to promoting a deeper understanding of Utah's diverse ecosystems and contributing to the sustainable, economic use of Utah's natural resources." Launched in 2010, an important first step for the consortium was to provide concise information on a publicly accessible website about each field station, including contact information, onsite resources, research focus, and links to the websites of the individual stations.[a] The UFSN website conveys an important educational message by emphasizing that gradients in elevation strongly influence the diverse natural ecosystems in Utah. Many of the individual field station websites also provide detailed information on the fauna and flora studied there.

The network has fostered integrated environmental research and education across the state. The educational value of connecting university and agency field stations is illustrated by the five-credit graduate field course that was organized in 2010. Students and three instructors followed a 16-day itinerary to field stations in three areas, learning field methods and meeting with resource managers. The topics included landscape patterns of vegetation, plant ecology and invasive species, range management, soils, microbes, wildlife, and natural history. It is now common for courses offered by one university to make use of multiple field sties in the network, and discounted user fees are granted to participating members.

[a] See http://www.utahfieldstations.org.

boundary focus common to each member's former national science agencies and build on the collective strengths of EU members' science capacity. Funding was provided only to collaborations that involved two or more EU countries. As a result, the collaborating countries' national science agencies expanded their horizons beyond national boundaries and started focusing on the collective strengths of every country's science capacity.

Universities and funding agencies could provide incentives for networking of field stations but should eschew top-down control. Funding agencies could encourage networking of field stations by giving preference to proposals that link multiple field stations. For example, the National Oceanic and Atmospheric Administration recently incentivized collaborations through its Office of Education in an Aquarium Initiative that restricts funding to proposals that involve two or more aquariums. The program resulted in several new collaborations among aquariums. Similarly, the National Science Foundation (NSF) Centers for Ocean Sciences Education Excellence also required collaborations between an informal education institution, a formal education institution, and a research organization.

As a first step in developing a bottom-up, voluntary, and effective field station network, one might focus on efforts to promote sharing data and information. Indeed, for field stations to become core components of a network, it is imperative that their rich repositories of data be made available to the broader scientific community in a timely way. An added challenge, however, is to develop protocols for collecting and aggregating long-term, place-based biological and environmental data in a uniform manner. Once long time-series data and information are collected and shared in a uniform manner across field station sites, field stations can collectively become major contributors to assessing ecological change at a larger scale and contribute to environmental resilience and sustainability science.

In addition, bottom-up, voluntary field station networks could replicate benefits offered by NSF-sponsored research coordinating networks, such as the LTER networks, which funds regional opportunities for field stations to develop shared research questions, allow graduate students and faculty to interact beyond their normal extent as colleagues, and share outreach and teaching programs. An important part of the network would extend beyond academe to include potential government partners—such as the US Fish and Wildlife Service, the National Park Service, the Bureau of Land Management, the U.S. Forest Service, state and local parks and forests, and nongovernment organizations such as The Nature Conservancy and land trusts (Box 3-3). When possible, representatives of private industry, foundations, agricultural research stations, and others might be involved in bringing research and real-world needs together.

Communities surrounding field stations are important stakeholders. Involving local communities in establishing and growing a network of users could benefit the long-term viability of field stations. As discussed in previous chapters, local communities have helped stations avoid closure, and could play a role in identifying and contributing to research on locally important issues. By including a variety of local, state, and federal agencies and land-management agencies, a

BOX 3-3
Partnerships with National Parks

State and national parks often have needs for research but insufficient research staff. Field stations and associated institutions can provide intellectual capital, but they need access to large, relatively undisturbed ecosystems, including terrestrial sites and state or federal marine protected areas. Field stations are in several U.S. national parks: Yosemite, Channel Islands, Sequoia, Kings Canyon, Haleakala, Capitol Reef, Cuyahoga Valley, Santa Rosa Island, Lassen, and Grand Teton (see, e.g., http://www.uwyo.edu/uwnps/). Field stations also operate in national parks in Costa Rica (the Organization for Tropical Studies in Palo Verde and Sirena in Corovado) and Ecuador (the Charles Darwin Research Station in the Galapagos). Field stations in the U.S. National Park Service (NPS) are affiliated with a wide variety of institutions and in some cases work with the 19 NPS research learning centers (RLCs, http://www.nature.nps.gov/ rlc/). RLCs are places where science and education come together to serve either one park or a network of parks. A standard policy at NPS sites requires data to be reported each year and in archived final reports. Products of the field stations include metrics on accommodations provided (person-nights) for researchers and students, peer-reviewed publications, reports based on park resources or managers' needs for information, and workshops, seminars, symposia, and partnerships with K-12 schools that involve thousands of students in programs related to science, technology, engineering, and mathematics. These activities can conflict with NPS goals and policies. However, all the field stations have been able eventually to develop working protocols to avoid such conflicts.

working group could develop strategies to share facilities, equipment, knowledge, and outreach capacity. Field stations could bring science to government land-management agencies, and in return, government land-management agencies could encourage research on government lands. Facilities and land are expensive and can often be shared for research and teaching. Interactions between the agencies and academic institutions could help answer important questions about real-world societal needs.

Developing networks at various levels will require new uses of resources and perhaps some reorganization and reallocation of existing resources. Energizing a "critical mass" of these institutions in collaborative observational programs could provide new insights into global change. For example, a network of field stations might offer data and information on regional damage, resilience, and recovery responses to extreme weather events, such as Superstorm Sandy, that would not be possible with data from a single station (Figure 3-2). Ideally, the networking would include terrestrial and marine field stations so that air–sea–land interactions could be studied and better understood. Developing networks will also require leadership at field stations to carefully evaluate their resources and policies (e.g., data management and sharing, tenure and promotion, and business development) to determine how to best incentivize collaborations and leverage each station's resources. Core sets of institutions could be established to demonstrate the power of networked observations, and additional networks could be developed in response to societal needs for Earth observations.

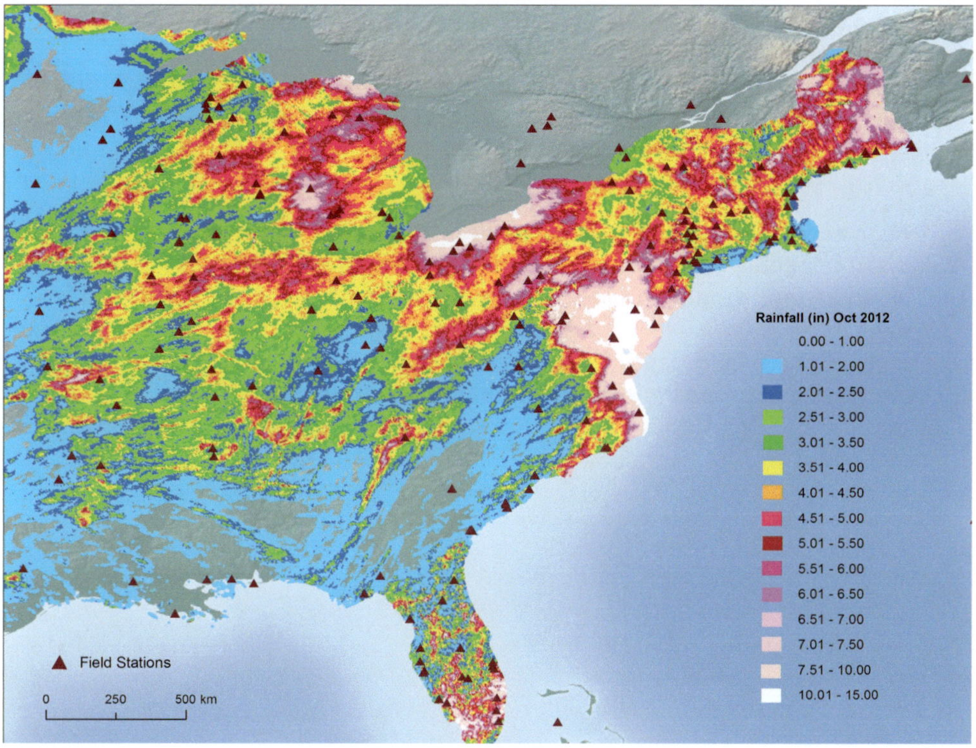

FIGURE 3-2 Field stations within the impact range of Superstorm Sandy. The map depicts the rainfall totals associated with Superstorm Sandy in October 2012. Red triangles indicate the locations of field stations.

Conclusions

Many field stations operate independently. Greater networking of the nation's field stations would offer many benefits and improve their ability to address important emerging environmental and societal issues. If the nation is to realize the benefits of its investment in field stations, the stations need to participate in multiple levels of networks and be integrated (at least virtually) into a nested set of interactive systems.

Greater networking would offer opportunities to save money, leverage resources, reduce redundancy, and increase effectiveness by sharing data and information, infrastructure, staff, and programs. Through networking, the field stations would reduce redundancy by sharing best practices, monitoring protocols, and platforms for data archiving and retrieval. Shared cyberinfrastructure will become increasingly important as the sizes of datasets grow. Sharing information and networking scientists could open new areas of scientific inquiry, education, and outreach. Networking can facilitate the development and diffusion of knowledge and technology in a way that nurtures innovation. It can capture social and intellectual capital to tackle major questions and seize opportunities as no single field station can, and it can enhance creativity and innovation by attracting the best people and promoting multidisciplinary collaboration. As a result, networking would improve the ability of participating field stations to document environmental change on a variety of scales both spatial and temporal. The most successful and sustainable networks start small and are self-defining; they encourage reciprocity among network members.

> ***Recommendation***: Field stations should seek opportunities for networking that make scientific, educational, and business sense. Universities and funding organizations should provide incentives for networking of field stations that meet those criteria. NSF and other funding agencies could encourage networking of field stations through the request-for-proposal process by giving preference to proposals that link multiple field stations.

4
Building and Maintaining a Modern Infrastructure

Efficient investment in scientific infrastructure requires long-term planning and clear and transparent decision making.
—UK House of Lords Science and Technology Committee, 2013

The infrastructure provided by field stations is essential to advance science in a rapidly changing world. The National Research Council's report *Critical Infrastructure for Ocean Research and Societal Needs in 2030* (NRC 2011) identifies next-generation categories of infrastructure that should be included in planning, provides advice on criteria that could be used to set priorities for asset development or replacement, recommends ways in which federal agencies could maximize the value of ocean infrastructure investments, and addresses societal issues. Because many parallels can be drawn between infrastructure needs in ocean research and those in field station–based research, the committee developed a modified definition of infrastructure on the basis of the 2011 NRC report (Box 4-1)

BOX 4-1
Definition of Infrastructure

Field station infrastructure is the full portfolio of resources and assets that include technology, facilities, data, people, and institutions that can be brought to bear in answering questions about Earth, the oceans, and the atmosphere and that are (or could be) shared by or accessible to the research community as a whole.

Field station infrastructure has two tiers:

- **Tier 1.** Field stations themselves as collective elements of a nation's broader scientific infrastructure.
- **Tier 2.** Individual components of field stations, such as laboratory space, scientific equipment, biological collections, cyberinfrastructure, historical data records, among others.

To ensure that field stations are adequately equipped to address and adapt to rapidly changing needs in science and education, consideration must be given to the organization and maintenance of both tiers of field station infrastructure. The question of how to maximize the value of field stations as components of a larger scientific infrastructure was addressed to a great extent in Chapter 3. The present chapter touches briefly on Tier 1 infrastructure but focuses primarily on Tier 2.

Recognizing the importance of science infrastructure, the European Union (EU) established the European Strategy Forum on Research Infrastructures (ESFRI) to enhance the use and management of large-scale and mid-scale research infrastructure and to facilitate scientists' access to research sites throughout the EU with the intent of strengthening its international reach (Figure 4-1). Eight large-scale facilities form the Partnership for European Environmental Research.

The U.S. National Science Foundation (NSF) maintains multiple programs that provide funding for science infrastructure. However, the United States does not have a central body that oversees scientific infrastructure, and it stands to learn from the ESFRI effort.

The die has been cast in part by the call in the National Research Council report on critical infrastructure (NRC 2011) for "a coordinated national strategic plan for critical shared ocean infrastructure investment, maintenance, and retirement." A similarly coordinated strategic plan is needed for field stations.

Defining Infrastructure Needs

There is no single list of infrastructure needs that fits all field stations. Field station infrastructure needs are driven by the strategic missions of the stations, the ecosystems within which they are embedded, the research questions they are addressing, and the levels of financial support they receive. Field stations vary along a continuum, from ones that have relatively simple infrastructural needs to those that have complex and sophisticated needs. The committee identified three basic types of field stations that reflect the continuum:

- Field stations that include little more than restricted access to research and teaching sites, parking, simple rustic housing or camping facilities, and a caretaker. These stations are used mainly for short-term visits by researchers that may recur over many years.
- Field stations that have laboratory space and housing, some autonomous environmental sensing equipment or data loggers, and an array of basic laboratory and field equipment, from microscopes and freezers to surveying equipment, small boats, and a support staff for maintenance. These stations often are used by researchers and classes for short- to intermediate-term stays.
- Field stations that have infrastructure resembling that of modern research laboratories that are engaged in cutting-edge science relevant to the study of ecosystems. They can incorporate a wide array of platforms (such as small and large boats and cyberinfrastructure), sensor networks, and other specialized facilities for accessing remote or extreme environments, including those in tree canopies, deep sediments, ice-covered habitats, and the open ocean; and they

have resident faculty and support personnel. These stations support a wide array of users, from resident researchers and site-based classes to day visitors and community events.

Every field station has strengths that make it appropriate and attractive for conducting particular kinds of research, education, and public outreach. That a field station has relatively simple infrastructure should not belie its value for research, teaching, or outreach. Indeed, the very nature of the site—its remote location, secure and rapid access to a particular ecosystem, and the absence of public disturbance—might be its greatest asset. The diversity and range of programs at field stations and their settings provide access to critical habitats, research opportunities on resident species, and sensor data (e.g., weather data and webcam videography) of interest and importance to local, regional, and national communities. Each field station's infrastructure should align with its vision and mission and the needs of its users.

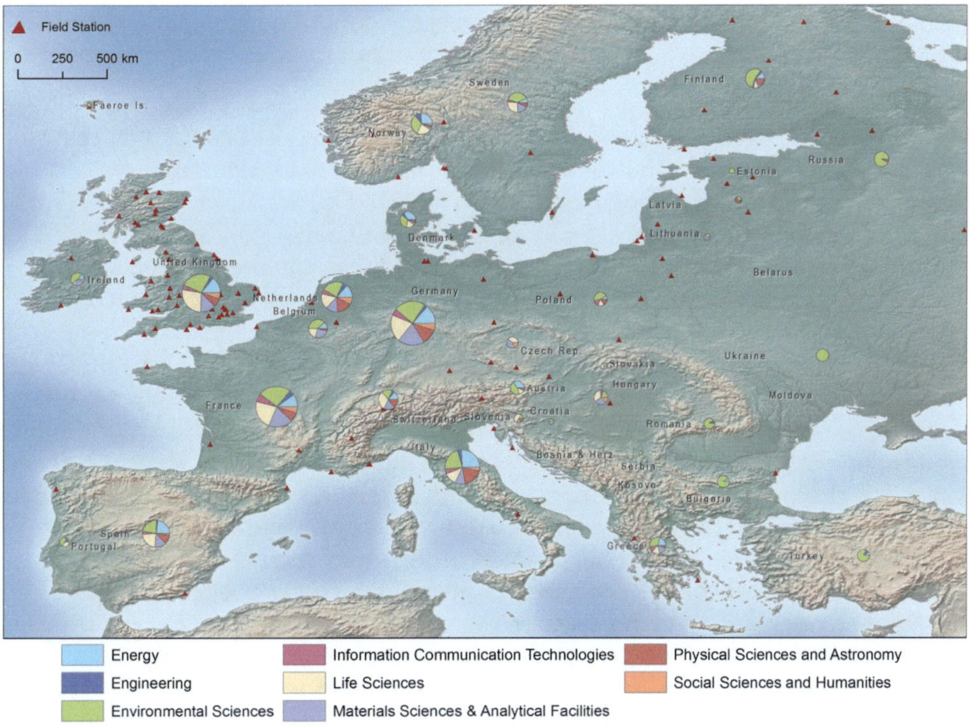

FIGURE 4-1. Large-scale research infrastructures funded by the European Union to provide transnational access to scientists across Europe. Map depicts country of and categories of research conducted at large-scale research centers (pie charts) overlain with the location of biological field stations (red triangles). The pie-chart diameters reflect the number of research facilities in each country (maxium = 139, Germany). Data to produce the pie charts were extracted from the website http://ec.europa.eu/research/infrastructures/index_en.cfm?pg=mapri, ©European Union, 2013.

Challenges of Maintaining and Upgrading Infrastructure

Field station managers and users have long recognized the need for safe, functional housing and properly equipped workspaces. Their two primary challenges in this regard are maintaining aging facilities and keeping up with rapid advances in technology. The latter is particularly important because laboratory-based research is increasingly integrated with field research. All infrastructure requires preventive maintenance, replacement, upgrading, or some combination of the three. That is not peculiar to field stations. However, the sites in which field stations are embedded and that make them attractive—along coasts, in mountains, in forests, or in deserts—often expose their infrastructure to extreme, highly variable environmental conditions that can take a toll. In addition, many stations are located in areas that are vulnerable to wildfires, earthquakes, tornados, hurricanes, or other natural hazards. These vulnerabilities add to the cost of maintaining the facilities and pose a risk to research equipment (e.g., laboratory equipment such as microscopes, autoclaves, and ultra-cold freezers, and field equipment such as nets, boats, and environmental sensors), biological collections, and data stored at these stations. Field station facilities can degrade much more rapidly than equipment found in environmentally controlled laboratories, and this is a financial burden on field station managers and in some cases compromises the research.

The recent survey by the National Association of Marine Laboratories (NAML) and the Organization of Biological Field Stations (OBFS) reveals common infrastructure priorities among field stations (NAML-OBFS 2013b). The top priorities are electricity, Internet access, support staff, laboratory space, storage, long-term monitoring, classroom capacity, housing, on-site maintenance, and engineering capacity. Respondents suggested that increased support for Internet access would improve scientists' ability to use field stations while providing potential visiting scientists with access to specific data catalogs that are critical for developing research programs. According to the survey, a major problem is that basic data catalogs—species lists, maps, weather data, and land-use history—often are lacking at field stations. Some respondents indicated that field stations had insufficient space for laboratories, classrooms, and storage (including refrigeration). Data-management systems were considered excellent by a few respondents but ineffective by others. In addition, field researchers may require transport to and from field sites. Transport needs can vary from a golf cart to a submersible, depending on the site and the research being conducted. Field stations with increasingly sophisticated scientific equipment and automated sensors will also have to make investments in the capacity to capture, process, store, and share increasingly large datasets. Consideration must also be given to data that do not typically lend themselves to classic deposition in accessible databases, such as video recordings of animal behavior or deep-sea observations. Upgrading data-management systems was also identified as a high priority in the survey.

Investments in maintaining existing infrastructure clearly are a primary concern for field station administrators but often are a relatively low priority for their host institutions, particularly if a field station is remote. Only 14 percent of respondents

($N = 197$) to the NAML-OBFS survey noted that financial planning for field stations included depreciation of buildings and equipment. That result is a remarkably low percentage, considering the respondents' overwhelming sense of vulnerability to anticipated funding losses in operational revenue (76%) and in federal (65%), state (60 %), administrative (54 %), and donor (54%) support over the next 5 years. There is clear need for every field station to develop a comprehensive infrastructure-management plan that is integrated with its strategic mission, its science plan, and its business plan (Lohr 2001).

Cyberinfrastructure and Connectivity

The inclusion of data as a type of infrastructure represents a paradigm shift for many field stations. Data constitute a primary product of field stations; if these data were made easily available, they could serve a broad audience. Long-term and baseline natural-history data should be an attraction to scientists and educators and be counted as part of a field station's value (see Chapter 6), and move them from serving merely as environmental sentinels to active participants in solving ecological and economic problems at a variety of scales. The acquisition of data is only one part of the equation. Data must be stored, managed, and integrated to ensure that they can be mined, visualized, and accessed through high-performance Internet connections—all parts of the domain of cyberinfrastructure.

Cyberinfrastructure consists of the assortment of information technologies that enable data storage, management, integration, and analysis. It is increasingly recognized as essential to science in that it dramatically improves scholarship and research productivity. Efficient cyberinfrastructure generally requires reliable Internet connectivity and modern computer hardware. At a minimum, field stations need adequate Internet connectivity to facilitate user access and collaboration. The availability of adequate cyberinfrastructure attracts scholars who are interested in cross-disciplinary research and fosters new scientific endeavors in emerging fields. Every field station should provide—whether on site, at a selected hub location (such as a host institution, the National Ecological Observatory Network (NEON), or other research centers), or through a collaborative network—online access to the complete historical datasets of its natural and human history and provide means by which its users can contribute to these datasets. This type of interactive access to databases can provide quality control of data in that scientists can monitor data input and output in real time and respond to anomalies. Scientists, students, or even visitors can see how data that they collect fit into larger temporal and spatial contexts.

Data Management

Infrastructure to organize, archive, and share data collected at a field station could expand the impact of a field station's research by making data available to other researchers to use, and by facilitating the ability to track data use and impact. Many tools for ecological data storage and recovery have been developed by the Long Term Ecological Reserve Network (LTER), the National Center for Ecological Analysis and Synthesis (NCEAS), the Knowledge Network for Biocomplexity (KNB), and others. Ecological metadata language developed by KNB and NCEAS has been widely used and is compatible with the larger aggregators (such as DataONE and Google). The National Park Service (NPS) has a research permitting and reporting database and a website that allows investigators to request reports and research data from specific national parks. The NPS website can be searched by park, taxon, or investigator.

Sharing the data products from field stations broadly would add value to the data and to the field station where the data were collected. Without centralized repositories, data developed at field stations are easily lost. Alternatively, if they are archived and made widely available, they have ever-increasing value to provide perspectives on environmental change. Archiving and sharing data from field stations are critical. The committee agrees with National Science Foundation's (NSF) current policy that data become publicly available after 2 years of completion of NSF-funded projects, and believes that field stations should adhere to this standard regardless of the funder.

Most institutions that fund research have a basic expectation that recipients will have specific data-management and data-sharing plans that will advance scientific objectives, maximize learning, and improve understanding of the outcomes of public investment by providing timely and long-term public access to, and relatively straightforward retrieval of, their data. With the shift from "small science" to "big science" (Meyer 2009) and the advent of large-scale, long-term interdisciplinary projects, such as LTER and NEON, collaborators grew to expect not only access to each other's data but data-management and data-sharing protocols built into the specific projects. That expectation was heightened in February 2013 when the Obama administration directed federal agencies to develop—in collaboration, if possible—plans to make federally funded research data freely available to the public within 1 year of publication as allowed by law.

The stricter guidelines for data management raise two critical questions: How are data to be stored? Who bears the cost? Some types of data (such as biological distribution data in spreadsheets) lend themselves to classical data-deposition methods, whereas others (such as video recordings) often do not. An example of the former is the data-management and data-sharing practices of VertNet,[32] a publicly accessible database of vertebrate distributions compiled by 86 institutions worldwide. The site is maintained by NSF and managed by a small staff, but the contributing institutions serve as the authoritative sources, providing and controlling the data that appear on the website. Exponentially increased exposure, use, and correction of the data result in higher data quality and greater intellectual exchange among participating researchers (Constable et al. 2010). This system incentivized collaboration and data sharing to great effect.

Typically, support for data management starts when the institutions provide research funds and persists only for the lifetime of the award. The continuing costs

[32] http://vertnet.org

of data management fall to the home institutions, which generally consider them to be fundamental to the conduct of research, preserving both research quality and academic integrity (see, e.g., the University of Oxford research data management website[33]). Universities that have extensive research activities can afford this approach, but it is unlikely that many small independent field stations can bear the additional economic burden of even the most basic data-management system.

The Dark-Data Problem

Researchers at field stations often record data in their logbooks and spreadsheets either by hand or electronically. They take the data with them and, historically, rarely share them with the field stations where they conducted their research. Some of the raw data are eventually analyzed and the results incorporated into peer-reviewed publications; some may be lost when a researcher is no longer active, and these fall into the realm of "dark data[34]"—data that are inaccessible to the broader scientific community that relies on new, more sophisticated data-management tools. Salvaging the large body of historical dark data that still reside in notebooks, file cabinets, or memories of aging investigators is a challenge, but worth pursuing.

The Berkeley Ecoinformatics Engine,[35] funded by the Keck Foundation at $3.5 million, could serve as a model for addressing both the dark-data challenge and the problem of integrating diverse databases within regional networks of field stations. The intent of the program was to organize and unify the wealth of data in University of California, Berkeley laboratories, natural-history museums, and field stations and to merge them with diverse environmental baselayers on climate, land cover or use, vegetation indexes, hydrology, and fire and other freely available datasets. The results are available for rapid exploratory analyses, tests for correlation, and visualizations that communicate results to a broad community of users. The Ecoinformatics Engine unites previously disconnected perspectives from Earth and atmospheric scientists, geographers, paleoecologists, and ecologists and enables tests of predictive models of global change. This constitutes a critical advance in making the science more rigorous.

Scaling Up and Sharing

Field station cyberinfrastructure is physically and technically diverse—from digital sensor arrays to high-speed communication networks—and varies widely among field stations. Because of the diversity, a comprehensive infrastructure-

---

[33]http://www.admin.ox.ac.uk/rdm

[34] Data that are not systematically indexed or stored in a manner that is accessible to the broader scientific community, such as biological specimen collections, analog data (e.g., observations recorded in laboratory notebooks), and data only found in research publications. Such data are "nearly invisible" and probably will be underused or lost (Heidorn 2008).

[35]http://ecoengine.berkeley.edu

management plan is best constructed around broad categories of *use* rather than *type* (physical, technical, and cyberinfrastructural). Three such use categories, modified from those described by ESFRI for large-scale research infrastructures (such as that of CERN, the European Organization for Nuclear Research), are (1) single sites, including infrastructure on the site of the field station itself; (2) networked sites, distributed resources and databases and infrastructure that are shared, possibly through collaborative networks; and (3) global infrastructure, available through online networks.

Those categories could be used to outline infrastructure in the context of individual field station needs *and* services. For example, the infrastructure-management plan would describe how and when data collected with a place-based infrastructure are to be stored (remain part of the field station infrastructure) and how and when they are to be shared in a distributed framework (a service provided by the field station to the scientific community).

Sharing information and resources among field stations is critical in a world in which technological advances and expenditures increase at a rapid rate. The resources may include datasets on soil types, land-use history, climate, and aspects of biology. Through networking, field stations and researchers can share resources and collaborate on common topics and scientific questions. Sharing of data requires use of standard formats and metadata.

When current best practices for data storage and metadata registry at the network level are used at a field station, they can become a part of a much larger, global infrastructure as modern data aggregators and information-management tools develop (e.g., DataONE and Google). As future information technology allows greater access to multiple data sources, the need increases for uniform data collection on target organisms and environmental properties and processes to allow analyses on regional and national scales.

Conclusions

Field stations vary in scope, size, and purpose; each contributes to the national research and education portfolio in critical ways. No array of infrastructure is applicable to all field stations, although there are similarities within each range of size and complexity. What is clear is that financial demands on field stations are increasing as they upgrade to meet today's science challenges. Installation of new cyberinfrastructure requires data-management and data-sharing plans and data that conform to widely used metadata standards. Such infrastructure requires a long-term commitment of experienced technical support. High-tech infrastructure generally has a relatively short life cycle (about 10 years), and provision for timely upgrading and replacement of any newly funded infrastructure is needed. Staff at many field stations, particularly smaller ones, do not have the required technical expertise.

Recommendation: Because of their wide variety in purpose size, and scope, each field station should assess and define its own infrastructure needs.

However, Internet connectivity and cyberinfrastructure should be included in all infrastructure-management plans to allow field stations to facilitate collaborative research and participate in broader networking efforts. The process of archiving dark data into digitally accessible formats is critical, and should begin with the most recent datasets and progress back in time so that field stations can expand their sets of continuous longitudinal data.

5

Strategies for Financial Sustainability

Don't put all your eggs in one basket.

—Proverb

Many field stations need a substantive transformation of their business practices and a long-term vision and strategy if they are to be financially sustainable. A report of the National Association of Marine Laboratories (NAML) and the Organization of Biological Field Stations (OBFS) and the details the future directions for science at field stations and provides some guidance to ensure that field stations are well positioned to advance research and innovation, education and training, and outreach in the 21st century (Billick et al. 2013). The NAML-OBFS report recommends that field stations increase their operational effectiveness, but it does not outline strategies on how to do it. In this chapter, the committee outlines how field stations should combine visionary leadership with strategic science and business planning to ensure long-term viability.

The goal of a comprehensive planning effort is to identify and articulate a compelling strategic vision and mission, to identify the assets of a given field station (or network of stations), to identify the future research challenges that it is uniquely positioned to address, to identify its unique education and outreach goals, metrics for measuring progress toward its goals, and to identify its value proposition (see Box 5-1). A programmatic planning effort should be supplemented with a business plan that makes explicit the field station's value proposition and that includes strategies that contribute to its financial sustainability.

In developing a business plan, a field station should start by identifying its assets, including the products and services that it provides. In the aggregate, these become key elements of the field station's value proposition and can lead to

BOX 5-1
Definition of Value Proposition

"[In marketing,] an innovation, service, or feature intended to make a company or product attractive to customers." (Oxford Dictionary)

A field station's value proposition can include elements as varied as longitudinal datasets, housing and conference facilities, extension and outreach learning opportunities for local communities, stewardardship of local natural history, and provision of rich research experiences at the convergence of disparate scientific disciplines. A field station's value proposition will be of interest to an array of stakeholders, including scientists, funding agencies, alumni, local or nearby business owners and communities, citizen scientists, and possibly major corporations.

revenue that adds to the core support providing a hedge against fluctuations in any one source of support. Potential assets that can be monetized include educational programs, room and board, personnel (such as technicians), access to laboratory equipment, biological collections, and even access to data that have been collected at a site and that provide the context for a visiting investigator's work. Each field station may have a different set of assets given its location, size and array of facilities, available long-term datasets, research equipment. personnel, and other resources. Assets that reflect the unique qualities of field stations, such as their physical location and access to distinctive ecosystems and their capability to merge science, education, and outreach unlike other institutions are particularly important for developing their value proposition. Executing the process identifying and assigning value to assets not only will allow a field station to inventory and document its assets, but will provide the information to market the assets to generate diverse sources of funding in support of its facilities and programs. However, careful consideration should be given to whether assets should be monetized for logistical, historical, or other strategic reasons. Successful expansion of public–private partnerships depends on a stable core of support that can be leveraged through a diversified value proposition that attracts an array of funding streams. It also depends on the visionary entrepreneurship of field station leaders who can attract and inspire funders and other partners.

Visionary Leadership

Effective leadership is one of the most critical factors in financial sustainability of field stations (Lohr 2001, NRC 2005). If field stations are to survive in the 21st century, they need leaders who will make them indispensable to their parent institutions. The idea of a field station as a separate and independent unit, often so remote as to be forgotten, is a thing of the past. The increasing sophistication of science, challenging economic realities, and the demand for greater accountability conspire to increase the demands and expectations of field station leadership. It behooves parent institutions and trustees to choose directors wisely and to place appropriate emphasis on the skill set a director will need to succeed.

There are parallels between the management needs of field stations and the management needs of businesses. One could say that field stations are in the "business" of scientific research, conservation, education, and public outreach. As such, the management of a field station requires individuals not only with scientific credentials, but also skills and experience in running successful businesses.

Often, the most common criterion in selecting the director of a field station, particularly one affiliated with a university, is the person's stature as a scholar who will command the respect of participating faculty. Equally important, however, is that the director have strong leadership and management skills and is able to gain the respect of employees, students, potential funders, and members of the public. The leader needs to be willing to put the success of the organization that he or she leads ahead of—or at least on a par with—his or her own success as a scholar. The metrics for assessing the director's performance should be clear and explicit before hiring. For example, if a leader is hired as a tenured or tenure-track faculty member at a university-supported field station, it is important that tenure criteria and performance metrics reflect the roles and mission of the field station, which may differ in part from the roles and responsibilities of on-campus faculty. Many of the skills needed for success in leading a field station are not part of the skill set of the typical academic scholar, and they may need to be enhanced or provided by another member of the field station's leadership/management team.

Field station directors are responsible both for building and sustaining the infrastructure that allows emerging science to thrive and for cultivating durable relationships with the parent organizations, the primary source of core support for most field stations. This requires that directors spend a significant amount of time working with key decision makers to develop a shared vision and trust while emphasizing the station's critical contributions to the parent institution's high-priority initiatives. This takes on particular importance for a field station affiliated with a university if it is to receive the same financial benefits and services as on-campus units. It also requires inviting university leaders and administrators to the field station, which is more easily accomplished once the relationship has developed.

Field station leadership may be implemented by using a variety of models. Depending on the size of the facility, various support staff may also be required to run and maintain it. In some cases, the dual roles—leader and manager of the field station—are held by one person, but in many cases the two roles are separated. The leader, or "champion," is likely to be a tenured academic who has a reputation as a researcher in a field relevant to the field station. The academic leader's salary often is a permanent line in a university budget. An operations manager, in contrast, is likely to be a person whose entire salary comes from the field station budget and may need more frequent justification to be established or maintained over the long term. A group of faculty that is committed to the success of the field station can often provide the energy and enthusiasm to sustain and support the leadership team and field station staff. Such a group can be vital in securing funds for research and for new infrastructure, in using the field station for classes, and in assisting in decision making and in securing institutional support. Finally, it is essential to plan for leadership succession. A change in a field station's director should not cause the business or programmatic underpinnings of a field station to falter. If a field station lacks adequate leadership, its long-term sustainability will be compromised.

Various models are available for leading and managing field stations. Data from the NAML-OBFS survey of field stations (227 responses from 444 potential field stations) indicate that 72% have station directors, 62% have maintenance staff, 51% have office staff, and 49% have research technicians (NAML-OBSF 2013b). Those were the most commonly reported staff positions. Much smaller percentages of the facilities have information technology staff (21%), research directors (21%), data managers (21%), or education staff (27%). These data are insufficient for developing the best performance model, which varies with the size, complexity, and management scheme, but they identify the models that are used most often.

Many directors will need training if they are to accomplish the leadership goals expected of them. Training is also critical for developing the next generation of effective science administrators. Training can be in the form of workshops that focus on creating vision and mission statements and business plans to support them, and that highlight successful leadership models that can be scaled to meet the needs of different kinds and sizes of field stations. Field stations associated with universities can work with their on-campus business schools in developing business plans. Those without business schools can turn to such organizations as the Senior Executive Service Corps that provide expert assistance to nonprofits at little or no cost to their clients. Workshops could be supported by NAML, OBFS, the National Science Foundation (NSF), or by other organizations. We applaud the recently launched Ecological Society of America Sustaining Biological Infrastructures initiative, funded by NSF.[36] Its first activity will be a workshop to train project directors in strategies for success. It is important to point out, however, that project management and program leadership and management are different challenges.

Success in a Time of Declining Resources

Funding from federal government grants and from most parent organizations, particularly universities, to support daily operations of field stations will continue to be a serious challenge for at least the near future. The bulk of support for most field stations generally comes from field stations' host institutions, but field stations have come to depend upon other sources to push their research and education agendas forward. Among them is the NSF program Improvements in Facilities, Communications, and Equipment at Biological Field Stations and Marine Laboratories. The program has provided more than $47 million in support over the last 15 years, specifically for the infrastructure needs of individual field stations at accredited U.S. universities and nonacademic organizations and should be continued (Figure 5-1 [37]). But the present financial situation of many field stations may not be sustainable. The committee believes that many field stations need to stabilize their base funding and diversify their funding portfolios.

If they are to be sustainable, many field stations need to make a more compelling case for their importance from the perspective not only of science and education, but also from other contributions they make to society if they are to be sustainable. Field stations can be important parts of local culture. They often maintain the best records of how the natural environment of a region has changed, and they may employ generations of local youth as field assistants or station workers. Evidence of a cognitive and physical health benefit from experiences in

---

[36]http://esa.org/sbi

[37]Information on the NSF awards was obtained from its public awards database, http://www.nsf.gov/awardsearch/.

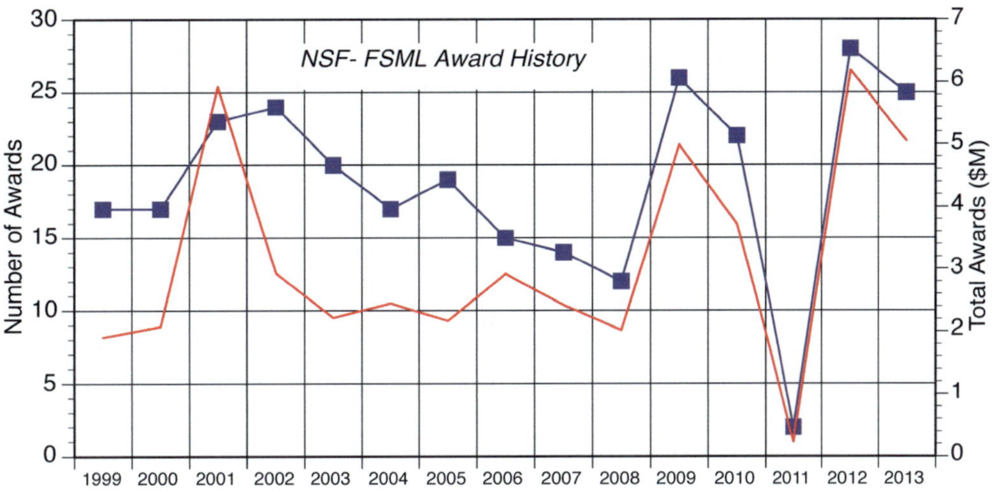

FIGURE 5-1. NSF field station and marine laboratory award history (1999-2013).The red line indicates the number of awards; the blue line indicates the total award amounts. The dip in 2011 reflects a change in the proposal deadline from early in the year to December. Data for this chart was obtained from the NSF public award database: http://www.nsf.gov/awardsearch/.

nature is growing (Bratman et al. 2012). In some nations, forests are being established as a form of medicine.[38] Field stations have a role to play in this growing sense of the major health benefits of interaction with nature. Field stations are also employers. For example, aquariums and marine laboratories in the Monterey Bay Crescent combined have 1,726 employees whose wages total more than $77.7 million (Miller 2007). Thus, thinking of the value of field stations simply in terms of scientific publications or even number of students taught yields too narrow a perspective. Field stations should make as broad a case as possible for the public good that they deliver.

A November 2013 report released by Secretary of the Interior Sally Jewell showed that national wildlife refuges contributed $2.4 billion to the U.S. economy and supported more than 35,000 jobs in 2013 (Carver and Caudill 2013). Field stations—particularly networks of field stations—would benefit from evaluating and sharing the links among their infrastructure and activities, stakeholder communities, and economic benefits (see Figure 5-2). Because a field station's infrastructure underlies all of its program and activities, a field station's value proposition, funding portfolio, and potential economic impacts are anchored in maintaining and upgrading its infrastructure. Economic impact analyses do not necessarily need to be conducted at each field station, but rather could be a coordinated effort of networks of field stations, to which economic multipliers can be applied for similar types of operations.

[38]http://infom.org

Field stations also generate indirect economic benefits, such as providing expertise and data for use by industry (such as commercial and recreational fishing, aquaculture, and renewable energy) and helping to create the next-generation workforce of scientists and technicians.

Return on Investment

Before a station's leaders develop strategies for long-term, sustainable support, they need to be able to answer a basic question: What is the return on investment?

Return on investment is a performance measure that is used to evaluate the efficacy of an investment. Field stations should be cognizant of the needs of their primary funding source (usually a university), local communities, and society at large and should actively and regularly seek their input. Knowing and understanding the community and societal needs can help field stations construct better and more effective research, education, and public-outreach programs and fundraising efforts. Appropriate metrics of the programs enable field stations to measure and articulate the returns on investment to current and future funders.

Stabilizing the Base

A stable, predictable, and adequate level of base support is a prerequisite for planning and is central to securing support from other sources to diversify a field station's funding portfolio. Stabilizing base funding support of field stations is essential, particularly for those affiliated with academic institutions. Universities and other funders not only should commit to a sustained level of base support for their field stations, but also need to provide professional financial and fundraising assistance to field station leaders, many of whom have had little experience in financial management and fundraising.

Continuity of support for field station infrastructure—including information technology, base maintenance and operations, and long-term operations—is essential for addressing our nation's environmental challenges in light of ever-increasing human pressures. Field stations should work together more effectively to share relevant data and other resources in a timely way. As discussed in Chapter 3, networking can make such efforts more efficient and effective.

Importance of a Diversified Funding Portfolio

One of the key factors in ensuring the stability and sustainability of field stations will be the development of diversified portfolios of funding sources that will be more resilient in challenging economic times. Diversity of funding sources reduces an institution's vulnerability to fluctuations in any one source of funding. Many field stations, particularly those affiliated with universities, depend too heavily on a single funding source for their support. As pointed out above, it is important that parent institutions provide a stable core of base support for their field stations, but they should expect that this support will be leveraged. Most field stations have many opportunities to diversify and supplement their sources of funding.

The first step in creating a diversified funding base is the development of a solid business plan. Many potential funders will insist on this before they will consider making an award. A business plan should also help in securing and stabilizing core support from a host university. The importance of this approach became clear to members of the committee in discussions with the former directors of the Wrigley Institute of the University of Southern California, the Hopkins Marine Station, the Mote Marine Laboratory, and the Duke University Marine Lab and with the present director of the Southern California Coastal Water Research Project.

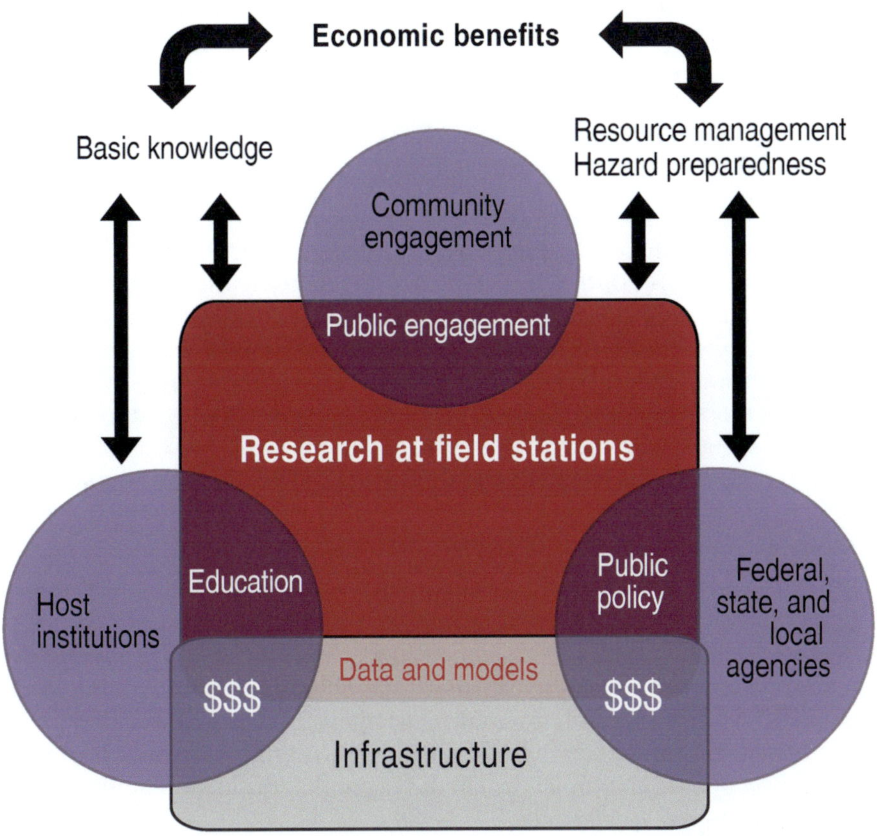

FIGURE 5-2. Links between field stations, stakeholder communities, and economic benefits. A field station's infrastructure underlies all of its programs. Infrastructure provides the basis for development of projects and activities in science, education, and public engagement along with connections to partners and stakeholders that share in these activities and related economic outcomes. Investments in constructing, maintaining, and upgrading a field station's infrastructure (buildings, equipment, biological collections, datasets, etc.) are linked to eventual economic benefits of field station activities.

Partnerships with private enterprises depend on financial motivations, such as research and development opportunities that have tangible profit outcomes or marketing relationships that may enhance a company's reputation or its products. Such opportunities may arise from time to time and should be seized when they occur and when they are appropriate, but they should not be counted on to make a large difference in the financial viability of field stations in general.

A networked community of collaborative field stations that shares resources, including human resources, will be more resilient than individual field stations in the face of stresses and shocks. A networked community will also be more capable of using technological advances to meet changing needs and to exploit new opportunities of science and society. The challenges and benefits of networking are explored in Chapter 3. Each field station will need to consider how to facilitate collaborations with other field stations and research organizations (e.g., for cost- and revenue-sharing reasons) that may operate with different funding models.

Conclusions

The value of field stations to society, local communities, and the nation warrants reliable institutional support. Aging infrastructure, the need for advanced technology and cyberinfrastructure, and evolving safety regulations are increasing financial demands on field stations as they upgrade to meet emerging science and societal challenges. Sustainable funding for modern infrastructure will be possible only if field station leaders develop compelling value propositions, strategic plans, and business models for operations that can secure base funding support which in turn can be leveraged by support from diverse sources. However, field stations leaders too often lack the required entrepreneurial skills. Effective business planning requires strong linkages to funding institutions and reaching out to diverse constituencies that can derive value from field stations.

> ***Recommendation***: Field stations and their host institutions should develop business plans that include clear value propositions and mechanisms to establish reliable base funding commitments that can be supplemented with funding from diverse sources. Business planning requires that station leaders be recruited not only for their scientific credentials but also for their leadership, management, and entrepreneurial skills. Host institutions should provide mentoring of field station leaders in management, business planning, and fundraising when appropriate.

6

Metrics for Achieving Goals and Demonstrating Impact

Counting sounds easy until we actually attempt it, and then we quickly discover that often we cannot recognize what we ought to count. Numbers are no substitute for clear definitions and not everything that can be counted counts.

—William Bruce Cameron, 1958

The value of field stations is widely documented in success stories by leading scientists, anecdotal evidence, and qualitative and semi-quantitative data. But it is difficult to analyze quantitatively the collective contribution of field stations to research, education, and outreach, because of the lack of aggregated empirical evidence. Field stations need common metrics that clearly demonstrate to their parent institutions and to current and future funders the range and magnitude of their impact. The few metrics that are available are haphazardly collected, fragmented, and infrequently shared. This weakens internal assessments and inhibits any synthetic assessment of the collective value of field stations to the scientific community and to broader society. The availability of the information in question is increasingly important as financial resources shrink or are reallocated to initiatives deemed to be of greater importance by funding institutions. In times of shrinking budgets, demonstrating outcomes and value become essential in securing long-term funding.

Key Elements for Developing Metrics

Sound metrics for evaluating program performance and progress include both quantitative and qualitative measures (NRC 2005). Traditional input metrics (number of staff or amount of research funding) and output metrics (number of publications, dissertations, and theses) alone paint an incomplete picture (NRC 2005). Outcome metrics, although more difficult to collect, are also important to assess the overall value of the field stations to science and society. Appropriate metrics are essential for monitoring, assessing, or modifying programmatic and financial strategies of field stations. Some of these key elements are highlighted in Box 6-1.

On the basis of those key categories, field station metrics should be developed to monitor and assess the impact and effectiveness of research, education, outreach, and financial strategies (see Box 6-2 for examples of such metrics). In addition, metrics should be meaningful to current and potential funders and to the

BOX 6-1
Key Elements for Developing Appropriate Metrics

1. Good leadership, governance, and strategic planning
2. Clear strategic plans that identify the core mission and articulate the goals against which progress can be measured
3. Robust business and funding plans to support infrastructure and research goals
4. Straightforward metrics that
   a. Encourage strategic assessments but avoid frequent, burdensome reporting
   b. Advance progress in research and education, are easily understood and accepted, and promote quality
   c. Maintain relevance, reflecting the dynamic and rapid pace of educational and scientific progress and objectives
5. Adequate human and financial resources for developing and applying useful metrics

SOURCE: NRC 2005 (pp. 3-4).

communities that field stations serve. Many field stations document at least one metric well, such as the number of research grants or number of publications, but fail to thoroughly document outputs and outcomes of other important activities, such as training, outreach, and achieving budgetary goals. Although it is essential to have metrics of standard performance, such as publications and grants, field stations also need to develop metrics to assess leadership success. For example, when measuring station director success, institutions should develop metrics to evaluate the leader's role in the success of the field station in carrying out its mission, not simply the director's career advancement as a scholar. Many smaller field stations lack the human and fiscal resources necessary for systematic gathering of data to document and archive metrics of performance. Thus, it is even more critical to have strong leadership with clear business and funding plans to enable efficient use of metrics with available resources.

Toward a Common Set of Metrics

The committee attempted to support the anecdotal evidence of the value of field stations to research, education, outreach, and career development with data and information on trends in funding, use, and impact. This proved to be an impossible task in the short time available and given the paucity of relevant information. Although each station might collect some data to demonstrate its contribution to research and education, the summative data and information for the broader community of field stations is neither stored nor accessible in a central location that we could identify. This is clearly a serious deficiency. We comment on some of the kinds of data that should be collected and made available so that, in the future, individual field stations and the community of field stations can make

---

BOX 6-2
Examples of Some Metrics to Assess Field Station Programs

Assessing impact of research

- Number of publications and their citation impact factors
- Number of digital datasets archived, downloaded, and cited
- Number of laws, regulations, and policies that have been influenced by field station research
- Participation in collaborative research and organizational networks

Assessing impact of education

- Alumni success stories
- Long-term tracking of field station students (e.g., graduation and career outcomes)
- Number of students conducting independent research (e.g., in Research Experience for Undergraduates and Experimental Program to Stimulate Competitive Research)
- Learning-outcomes assessments

Assessing impact of outreach

- Number of organizations that visit the field station (e.g., in summer programs and community organizations and through citizen science)
- Learning-outcomes assessments
- Media reach evaluation

Assessing field station use (research, teaching, and outreach)

- Number of user days and contact hours
- Peak season of use and capacity for facility

Assessing financial stability

- Number and size of grants enabled by field stations
- Amount of recovered overhead
- Revenue income from endowments, gifts, sponsored activities, and user charges
- Operating and maintenance expenses

---

a more compelling case of their value to science and society.

To make the case for their collective importance as a national and even international resource, field stations would benefit greatly from working together to develop a common set of metrics of performance to document their outputs and outcomes and to allow comparisons among stations. A common system for a wide range of metrics is important, so that evidence and trends of impact can be aggregated and differentiated across the wide range of missions and goals of individual field stations. Sharing the development and collection of metrics would be greatly facilitated if field stations were part of a network. This also would make an assessment of impacts less costly for individual field stations.. As outlined in Chapter 1, these metrics, data, and metadata will be an increasingly important resource for individual field stations, networks of field stations, the nation, and the world in this era of climate change and other environmental and societal pressures, as well as declining funding and increasing demand for accountability. If metrics are to be diagnostic, they will have to be scalable by station size and mission. For example, not all field stations place equal priority on teaching, research, and outreach, and as has been observed, they range greatly in size.

There is a need for a centralized capacity to store, manage, and distribute data on metrics. To accomplish those goals on a broad scale, the National Science Foundation (NSF) could support the development and implementation of a centralized database of field station metrics in collaboration with such professional societies as the Organization of Biological Field Stations and the National Association of Marine Laboratories. That would require an investment, but it would provide major benefits in documenting the contributions of individual field stations and the community of field stations, and in evaluating the relative contributions of different field stations, and identifying potential networking opportunities among facilities.

Measuring Progress and Impact

In general, measuring the impact of research at field stations on the scientific enterprise poses a challenge because benefits to society usually are not observed until years after research is completed. A few field stations are documenting metrics of research output: La Selva Biological Station produced over 3,000 publications from 1956 to 2007 (Michener et al. 2009), and the Rocky Mountain Biological Laboratory (RMBL) produces an average of 35 scientific publications per year (Billick and Price 2010); 1,324 publications and 97 dissertations had been based on research conducted at RMBL between its inception in 1928 and 2011 (Inouye 2013). However, tracking publications from field stations can be difficult. Modern data aggregators (e.g., altmetrics[39] and Google Scholar) could make publication tracking easier. For example, the University of Alaska of the North recently generated a Google Scholar profile (UAM Birds), to better assess the number and quality of publications supported by the museum's bird collection. The effort led museum staff to discover that "the body of work supported by the collection is diverse and well cited, with a profile h-index[40] of 42, equivalent to an average Nobel laureate in physics" (Winker and Withrow 2014; Figure 6-1). A field station–specific digital object identifier (DOI) would be even more advantageous for publication tracking: each field station could publish a basic description of its location (where it is and general characteristics of the site) and then submit the description to a stable ecological archive to generate a DOI. If each future publication based on research at a particular field station cited this basic description, including the DOI, publications from the field station could be easily

[39] http://altmetrics.org/manifesto

[40] Measure of a "scholar's" impact based on the number of publications and the number of citations per publication.

UAM Birds

University of Alaska Museum, Department of Ornithology
The science of birds
Verified email at alaska.edu

Citation indices

	All	Since 2009
Citations	7988	3764
h-index	45	33
i10-index	129	97

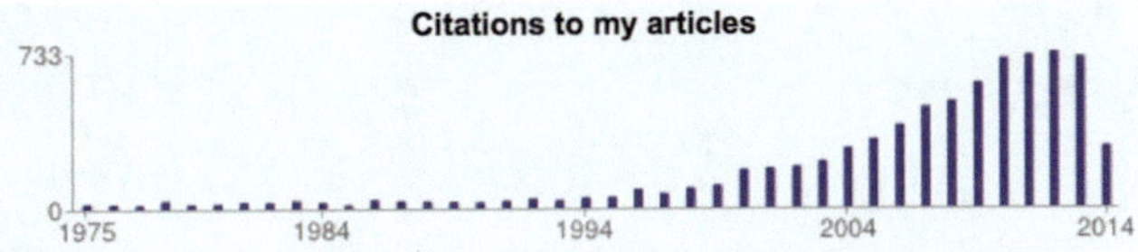

FIGURE 6-1. Google Scholar Page of the University of Alaska Museum Bird Collection. The scholar page lists all cited publications (not included here), tracks the number of publications over time, and calculates indices of citation impact.

tracked. That would not allow tracking of past activities, but it would allow tracking of future publications by field stations. The combination of a DOI and modern data aggregators could further facilitate the tracking of future publications on field station research.

DOIs could also be applied to a field station's raw datasets, metadata, or biological collections. Sharing the data products from field stations broadly would add value to the data and to the field station where the data were collected. A potential metric of the impact and use of data from field stations could be developed by analyzing how often metadata and datasets are downloaded, used, or cited. Outputs such as the number of research publications or frequency of dataset use are important, but they are not good indicators of outcomes or impact. Outcomes also require attention.

Societal impact could be measured by aggregating the number of laws, regulations, policy decisions, or global assessments that have been informed by field station research and data (see, e.g., Table 1-2). Aggregation could be handled by the proposed field station network or a third-party oversight organization. The network could also survey educators in the K-12 and university systems to assess what teachers' needs are in science, technology, engineering, and mathematics curricula and how field station programs have benefited or could potentially benefit teachers and students who visit. A network could prepare a report on the number of studies conducted or the number of researchers that use field stations each year and share this information with local communities, host institutions, government agencies, and other funders.

A variety of field stations have programs that can serve as models because they provide data repositories, libraries, lists of publications, information about the local climate, species found in the area, and scientists and staff members who work there. For example, the website of Archbold Biological Station includes annual reports, data and metadata, and even a fact sheet describing the location, habitats, climate, and species. The website for Cedar Point Biological Station, a smaller field

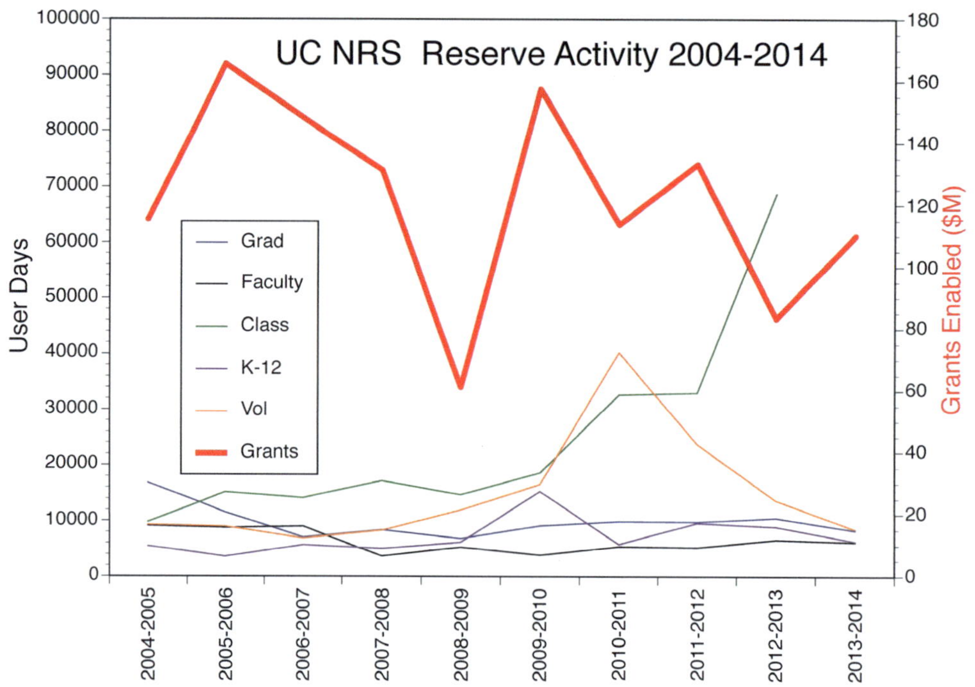

FIGURE 6-2. Aggregated Data on User Activities within the UC Natural Reserve System. User-reported data on field station user-days (1 day and generally 1 night) between 2004 and 2014 in self-defined categories as follows: graduate students (Grad), research faculty (Faculty), university-level students (Class); K-12 students; volunteers (Vol); and the total budgets of research grants that were approved to use one more of the UC NRS reserves (Grants). Prior to 2012, undergraduate and graduate student user-days were lumped together. Data for this chart was gathered from the Reserve Application Management System: http://rams.ucnrs.org/.

station that is associated with the University of Nebraska, provides information about science camps, facilities, and local natural areas and a link to alumni so that they can stay connected. The University of California Natural Reserve System uses centralized data collection to support its activities and is one of the first such databases[41] that makes reserve-related research widely available within a network (Figure 6-2; Box 6-3).

These examples show that it is possible to create simple, yet sophisticated systems for collecting data and information on field stations that show the value of field stations with respect to meeting their core goals. It may be pertinent to point out that private field stations often have a stronger Web presence than many public field stations. One reason for that may be that private stations need to fund their continued operations through private contributions and user fees and have found that a strong Web presence leads to more financial support. Field stations could learn from the Conservation Measures Partnership, which, with funding from several private foundations interested in improving organization effectiveness,

[41] http://www.ucnrs.org/bibliography.lasso

BOX 6-3
University of California Natural Reserve System—Collecting and Aggregating Data

The University of California Natural Reserve System (UC NRS) needed a way to track their activities so that they could report quantitative measures of their use to supporting campus administrators and external private sector and state and federal agency funders. However, gathering data on the 39 field stations within UC NRS across a wide variety of uses for research, education, and public engagement is complex. The UC NRS stations are visited by many different people, and provide infrastructure for multiple research projects that can span different time periods and include more than one reserve.

In 2000, the Reserve Application Management System (RAMS)[42] was created to address the challenge of gathering data on UC NRS use. RAMS captures information *from the users* of the field stations by asking them to fill out an application before they are allowed access to the reserves. The application gleans input data on every approved research project for inclusion in the core ecological metadata. RAMS also requires a user's research permit information and provides a liability waiver form online. Some NRS reserves were slow to adopt RAMS, and so underreporting is an issue. Nonetheless, RAMS enables station managers and UC NRS leadership to aggregate a wide-variety of quantitative data about station uses.

For example, from January 2010 to January 2013, 26,600 people spent a total of 84,237 user-days on the 39 reserves, including over 2,500 university-level researchers. More than 150 undergraduate courses were offered at one or more NRS reserves, including 3,900 university students. Over 1,700 K-12 students participated in learning on the reserves. Research activity on the reserves resulted in 683 peer-reviewed journal articles, books, and book chapters. Research grants enabled by the reserves totaled $386.4 million. Research projects enabled by more than one reserve accounted for $74.6 million of the total extramural grant funding.

In 2012, RAMS was upgraded to a MySQL relational database. RAMS metadata follow the Morpho format developed by the Knowledge Network for Biocomplexity[43] (see Michener and Jones 2012). The data is accessible online[44] and includes location, temporal span, abstract of research, author, contact information, and funding sources and amounts.

developed a global standard for conservation projects.[45]

Conclusions

The value of field stations is widely acknowledged but unevenly documented by scientists and station managers in anecdotal evidence and in qualitative and semi-quantitative form. Measures of effectiveness that are aligned with field stations' science, education, and business plans can lead to improvement in performance and impact but typically are lacking. In the absence of metrics, it is impossible to manage for improved outcomes.

---

[42] http://rams.ucnrs.org
[43] www.ecoinformatics.org
[44] http://rams.ucnrs.org
[45] http://www.conservationmeasures.org

Effective practices involve supporting and training leaders, collection of metrics, and networking among field stations. It is essential that all field stations have effective leadership and a strong support base that includes scientists, donors, and stakeholders that extend beyond the field station. It also is essential that field stations collect data that can be transformed into at least a minimal number of metrics to document their performance and successes. A strong communication program is necessary for both leaders and the institution and should include an effective and current Web presence. Finally, many functions of field stations could be enhanced by the formation of partnerships and networks, both nationally and internationally.

> ***Recommendation***: Field stations should work together to develop a common set of metrics of performance and impact. The metrics should be designed so that they can be aggregated for regions and the entire nation. Universities and other host institutions and funding organizations should support the gathering and transparent reporting of field station performance metrics because such information will enhance the stations' ability to document the contributions of field stations to the nation's research and education enterprise.

> ***Recommendation***: New mechanisms and funding need to be developed to collect, aggregate, and synthesize performance data for field stations, and to translate these data into metrics and information that can be used to document the value of the community of field stations to science and society.

References

AAAS (American Association for the Advancement of Science). 2014. What We Know: The Reality, Risks and Response to Climate Change. Available: http://whatweknow.aaas.org/wp-content/uploads/2014/03/AAAS-What-We-Know.pdf [accessed April 9, 2014].

Alper, J. 1998. Smithsonian field station gets the boot. Science 280(5368):1340a.

American Academy of Arts & Sciences. 2013. ARISE 2: Unleashing America's Research & Innovation Enterprise. Cambridge, MA: American Academy of Arts & Sciences. Available: http://www.amacad.org/content/publications/publication.aspx?d=1138 [accessed February 25, 2014].

American Institutes for Research. 2005. Effects of Outdoor Education Programs for Children in California. Available: http://www.air.org/sites/default/files/downloads/report/Outdoorschoolreport_0.pdf [accessed May 6, 2014].

Annand, V. 2014. Scientist to speak at AU laments Meanook research station's closure. Athabasca Advocate, April 15. Available: http://www.athabascaadvocate.com/article/20140415/ATH0801/304159982/0/ath11 [accessed April 17, 2014].

Barry, J.P., C.H. Baxter, R.D. Sagarin, and S.E. Gilman. 1995. Climate-related, long-term faunal changes in a California Rocky Intertidal Community. Science 267(5198):672-675.

BBC. 2013. Cumbrae Marine Research Centre under threat of closure. BBC Scotland, January 28. Available: http://www.bbc.com/news/uk-scotland-21229839 [accessed April 10, 2014].

Billick, I., and M.V. Price. 2010. The Ecology of Place: Contributions of Place-Based Research to Ecological Understanding. Chicago, IL: University of Chicago Press.

Billick, I., I. Babb, B. Kloeppel, J.C. Leong, J. Hodder, J. Sanders, and H. Swain. 2013. Field Stations and Marine Laboratories of the Future: A Strategic Vision. National Association of Marine Laboratories and Organization of Biological Field Stations. Available: http://www.obfs.org/assets/docs/fsml_final_report.pdf [accessed April 10, 2014].

Borgatti, S.P., and P.C. Foster. 2003. The network paradigm in organizational research: A review and typology. Journal of Management 29(6):991-1013.

Bowser, A., and L. Shanley. 2013. New Visions in Citizen Science. Woodrow Wilson International Center for Scholars, Washington, DC. Available: http://www.wilsoncenter.org/sites/default/files/NewVisionsInCitizenScience.pdf [accessed May 6, 2014].

Bratman, G.N., J.P. Hamilton, and G.C. Daily. 2012. The impacts of nature experience on human cognitive function and mental health. Annals of the New York Academy of Sciences 1249:118-136.

Brooke, R.K., and S.M. McLachlan. 2008. Trends and prospects for local knowledge in ecological and conservation research and monitoring. Biodiversity and Conservation 17:3501-3512.

Brossard, D., and B.V. Lewenstein. 2010. A critical appraisal of models of public understanding of science: Using practice to inform theory. Pp 11-39 in Communicating Science: New Agendas in Communication, L. Kahlor and P.A. Stoud, eds. New York: Routledge.

Cameron W.B. 1958. Tell me not in mournful numbers. National Education Association Journal 47(3): 173

Carver, E., and J. Caudill. 2013. Banking on Nature: The Economic Benefits to Local Communities of National Wildlife Refuge Visitation. U.S. Fish and Wildlife Service, Washington, DC. Available: http://www.doi.gov/news/loader.cfm?csModule=security/getfile&pageid=380921 [accessed May 6, 2014].

CBC News. 2011. UManitoba shuts down Delta Marsh field station. November 25. Available: http://www.cbc.ca/news/canada/manitoba/umanitoba-shuts-down-delta-marsh-field station-1.1086746 [accessed April 10, 2014].

CBC News. 2014. New vision for Ontario's Experimental Lakes Area unveiled. June 24. Available: http://www.cbc.ca/news/canada/kitchener-waterloo/new-vision-for-ontario-s-experimental-lakes-area-unveiled-1.2685764 [accessed April 10, 2014].

CENS (Center for Embedded Networked Sensors). 2013. Hopkins Marine Station: History. Stanford University. Available: http://www-marine.stanford.edu/history.htm [accessed December 28, 2013].

Conrad, C.C., and K.G. Hilchey. 2011. A review of citizen science and community-based environmental monitoring: Issues and opportunities. Environmental Monitoring and Assessment 176(1-4):273-291.

Constable, H., R. Guralnick, J. Wieczorek, C. Spencer, and A.T. Peterson. 2010. VertNet: A new model for biodiversity data sharing. PLoS Biology 8(2):e1000309.

Cruz, C. 2012. Pacific Biosciences Research Center wins reprieve. KITV.com. Available: http://www.kitv.com/news/hawaii/Pacific-Biosciences-Research-Center-wins-reprieve/17380764 [accessed April 9, 2014].

Darwin, C.R. 1887. Life and Letters of Charles Darwin, Including an Autobiographical Chapter. Vol 1. F. Darwin, ed., William Clowes and Sons, Ltd London.

Dirks, C. 2011. The current status and future direction of biology education research. Commissioned paper. Presentation at the Second Meeting on the

Status, Contributions, and Future Directions of Discipline-Based Education Research, October 18-19, 2010, Washington, DC. Available: http://sites.nationalacademies.org/dbasse/bose/dbasse_080124 [accessed May 7, 2014].

Dolan, J. 2007. On Kofoid's trail: Marine biological laboratories in Europe and their life histories. Limnology and Oceanography Bulletin 16:73-76.

EPA (U.S. Environmental Protection Agency). 2012. Science for a Sustainable Future. EPA 601/R 12/002. Office of Research and Development. Available: http://www.epa.gov/research/docs/strap-overview2012.pdf [accessed April 11, 2014].

Ewel, K.C. 2001. Natural resources management: The need for interdisciplinary collaboration. Ecosystems 4(8):716-722.

Feener, D.H., Jr. 1981. Competition between ant species: Outcome controlled by parasitic flies. Science 214:815-817.

Fischoff, B. 2013. The sciences of science communication. Proceedings of the National Academy of Sciencesof the United States of America. 110(Suppl. 3):14033-14039.

Fuller, B.1975. Everything I Know [Lecture]. Buckminster Fuller Institute. Available: http://bfi.org/about-fuller/resources/everything-i-know [accessed April 9, 2014]

Globe and Mail. 2013. Closing of research stations belies Ottawa's claim that it is protecting the environment [editorial]. The Globe and Mail, March 19. Available: http://www.theglobeandmail.com/commentary/editorials/closing-of-research-stations-belies-ottawas-claim-that-it-is-protecting-the-environment/article9940694/ [accessed April 9, 2014].

Graham, M.J., J. Frederick, A. Byars-Winston, A.B. Hunter, and J. Handelsman. 2013. Increasing persistence of college students in STEM. Science 341(6153):1455-1456.

Hake, R.R. 1998. Interactive-engagement versus traditional methods: A six-thousand student survey of mechanics test data for introductory physics courses. American Journal of Physics 66(1):64-74.

Hansen, J., P. Kharecha, M. Sato, V. Masson-Delmott, F. Ackerman, D.J. Beerling, P.J. Hearty, O. Hoegh-Guldberg, S. Hsu, C. Parmesan, J. Rockstrom, E.J. Rohling, J. Sachs, P. Smith, K. Steffen, L. Van Susteren, K. von Schuckmann, J.C. Zachos. 2013. Assessing "dangerous climate change": Required reduction of carbon emissions to protect young people, future generations and nature. PLOS ONE 8:e81468.

Harley, D. 2013. Scholarly communication: Cultural contexts, evolving models. Science 342(6154):80-82.

Heidorn, P.B. 2008. Shedding light on the dark data in the long tail of science. Library Trends 57(2):280-299.

Hewatt, W.G. 1934. Ecological Studies on Selected Marine Intertidal Communities of Monterey Bay. Dissertation, Stanford University, Stanford, CA.

Hewatt, W.G. 1937. Ecological studies on selected marine intertidal communities of Monterey Bay, California. American Midland Naturalist 18(2):161-206.

Hoag, H. 2013. Ecology: Lady of the lakes. Nature. October 20. Available: http://www.nature.com/news/ecology-lady-of-the-lakes-1.14043 [Accessed April 10, 2014].

Inouye, D. 2013. Long-term place-based contributions to ecology: The Rocky Mountain Biological Laboratory. OOS 11-4. Presentation at the 98th Ecological Society of America Annual Meeting, August 6, Minneapolis, MN.

IPCC (Intergovernmental Panel on Climate Change). 2001. Climate Change 2001: Impacts, Adaptation and Vulnerability. Cambridge, UK: Cambridge University Press. Available: http://www.grida.no/publications/other/ipcc_tar/?src=/climate/ipcc_tar/wg2/index.htm [accessed April 11, 2014].

IPCC. 2007. Climate Change 2007: Impacts, Adaptation and Vulnerability: Contribution of Working Group II to the Fourth Assessment Report of the IPCC, M.L. Parry, O.F. Canziani, J.P. Palutikof, P.J. van der Linden, and C.E. Hanson, eds. Cambridge, UK: Cambridge University Press. Available: http://www.ipcc-wg2.gov/publications/AR4/index.html [accessed April 11, 2014].

IPCC. 2014. Climate Change 2014: Impacts, Adaptation and Vulnerability. Available: http://www.ipcc.ch/report/ar5/wg2/ [accessed April 11, 2014].

Jack, H.A. 1945. Biological field stations of the world. Chronica Botanica 9(1):1-38.

Johnson, M.A., and A.E. Lawson. 1998. What are the relative effects of reasoning ability and prior knowledge on biology achievement in expository and inquiry classes? Journal of Research in Science Teaching 35(1):89-103.

Kalani, N. 2012. UH Manoa Faculty: No confidence in vice chancellor for research. Honolulu Civil Beat, March 21. Available: http://www.civilbeat.com/articles/2012/03/21/15290-uh-manoa-faculty-no-confidence-in-vice-chancellor-for-research/[accessed April 9, 2014].

Lohr, S. 2001. An Operations Manual for Field Stations and Marine Laboratories. Organization of Biological Field Station. Available: http://www.obfs.org/assets/docs/operation-manual/obfs-adminhandbook.pdf [accessed April 10, 2014].

Lopatto, D. 2007. Undergraduate research experiences support science career decisions and active learning. CBE Life Science Education 6(4):297-306.

Meyer, E.T. 2009. Moving from small science to big science: Social and organizational impediments to large scale data sharing. Pp. 147-159 in E-research: Transformation in Scholarly Practice, N. Jankowski, ed. New York: Routledge.

Michael, J. 2006. Where's the evidence that active learning works? Advances in Physiology Education 30:159-167.

Michener, W.K., K.L. Bildstein, A. McKee, R.R. Parmenter, W.W. Hargrove, D. McClearn, and M. Stromberg. 2009. Biological field stations: Research legacies and sites for serendipity. BioScience 59(4):300-310.

Michener, W.K., and M.B. Jones. 2012. Ecoinformatics: Supporting ecology as a data-intensive science. Trends in Ecology and Evolution 27(2):85-93.

Millennium Ecosystem Assessment. 2005. Ecosystems and Human Well-Being: Desertification Synthesis. Washington, DC: World Resources Institute. Available: http://www.millenniumassessment.org/documents/document.355.aspx.pdf [accessed April 10, 2014].

Miller, N. 2007. The Value of Marine Institutions: An Economic Survey of Marine Research and Education Institutions in Monterey and Santa Cruz Counties. Available: http://www.mbari.org/education/internship/07interns/papers/InternPapers/NMiller.pdf [accessed April 10, 2014].

MLBS (Mountain Lake Biological Station). 2014. ArtLab. Mountain Lake Biological Station, University of Virginia. Available: http://mlbs.org/artlab [accessed April 11, 2014].

Nadkarni, N.M., and A.W. Stasch. 2012. How broad are our broader impacts? An analysis of the National Science Foundation's Ecosystems Studies Program and the Broader Impacts requirement. Frontiers in Ecology and the Environment 11(1):13-19.

Nagda, B.A., S.R. Gregerman, J. Jonides, W. von Hippel, and J.S. Lerner. 1998. Undergraduate student faculty research partnerships affect student retention. Review of Higher Education 22(1):55-72.

NAML-OBFS (National Association of Marine Laboratories and Organization of Biological Field Stations). 2013a. Building and Operating Field Station and Marine Laboratories of the Future: Workshop Report, November 17-18, 2011. Available: http://www.obfs.org/assets/docs/fsml_workshop_report.pdf [accessed April 10, 2014].

NAML-OBFS. 2013b. Place-Based Research Site Strategic Planning Survey: Results Summary. Available: http://www.obfs.org/assets/docs/fsml_survey_report.pdf [accessed April 10, 2014].

NRC (National Research Council). 2001. Grand Challenges in Environmental Sciences. Washington, DC: National Academy Press.

NRC. 2005. Thinking Strategically: The Appropriate Use of Metrics for the Climate Change Science Program. Washington, DC: The National Academies Press.

NRC. 2009a. Learning Science in Informal Environments—People, Places, and Pursuits. Washington, DC: The National Academies Press.

NRC. 2009b. Nutrient Control Actions for Improving Water Quality in the Mississippi River Basin and Northern Gulf of Mexico. Washington, DC: The National Academies Press.

NRC. 2010a. Adapting to the Impacts of Climate Change. Washington, DC: The National Academies Press.

NRC. 2010b. Landscapes on the Edge: New Horizons for Research on the Earth's Surface. Washington, DC: The National Academies Press.

NRC. 2011. Critical Infrastructure for Ocean Research and Societal Needs in 2030. Washington, DC: The National Academies Press.

NRC. 2013. Abrupt Impacts of Climate Change: Anticipating Surprise. Washington, DC: The National Academies Press.

NRC. 2014a. Convergence: Facilitating Transdisciplinary Integration of Life Science, Physical Science, Engineering, and Beyond. Washington, DC: The National Academies Press.

NRC. 2014b. Sustainable Infrastructures for Life Science Communication: Workshop Summary. Washington, DC: The National Academies Press.

NSF (National Science Foundation). 2009. Transitions and Tipping Points in Complex Environmental Systems. A Report by the NSF Advisory Committee for Environmental Research and Education. Available: http://www.nsf.gov/geo/ere/ereweb/ac-ere/nsf6895_ere_report_090809.pdf [accessed April 11, 2014].

NSF. 2013a. Experimental Program to Stimulate Competitive Research (EPSCoR). Available: http://www.nsf.gov/od/iia/programs/epscor/nsf_oiia_epscor_index.jsp [accessed December 2013].

NSF. 2013b. Research Experience for Undergraduate (REU). Search Results for REU Sites. Available: http://www.nsf.gov/crssprgm/reu/list_result.jsp [accessed December 2013].

Orihel, D., and D. Schindler. 2014. Experimental Lakes Area is saved, but it's a bittersweet victory for science. Globe and Mail. April 1. Available: http://www.theglobeandmail.com/globe-debate/experimental-lakes-area-is-saved-but-its-a-bittersweet-victory-for-science/article17753956/ [Accessed April 10, 2014].

OSU (Oregon State University). 2014. Life-long Marine Science Learning. Available: http://seagrant.oregonstate.edu/education[accessed April 11, 2014].

Parmesan, C. 2007. Influence of species, latitudes and methodologies on estimates of phenological response to globa1 warming. Global Change Biology 13(9):1860-1872.

Parmesan, C., and G. Yohe. 2003. A globally coherent fingerprint of climate change impacts across natural systems. Nature 421(6918):37-42.

PCAST (President's Council of Advisors on Science and Technology). 2012. Engage to Excel: Producing One Million Additional College Graduates with Degrees in Science, Technology, Engineering and Mathematics. Available: http://www.whitehouse.gov/sites/default/files/microsites/ostp/pcast-engage-to-excel-final_feb.pdf [accessed May 6, 2014].

Pennisi, E. 2008. Hawaii Marine Lab fights to stay afloat. Science 322(5900):358.

Pentland, A. 2014. Social Physics: How Good Ideas Spread—The Lessons from a New Science. New York: Penguin Press.

Roco, M.C., W.S. Bainbridge, B. Tonn, and G. Whitesides, eds. 2013. Convergence of Knowledge, Technology, and Society: Beyond Convergence of Nano-Bio-Info-Cognitive Technologies. New York: Springer. Available: http://www.wtec.org/NBIC2/Docs/FinalReport/Pdf-secured/NBIC2-FinalReport-WTECversion--web.pdf [accessed February 26, 2014].

Root, T.L., J.T. Price, K.R. Hall, S.H. Schneider, C. Rosenzweig, and J.A. Pounds. 2003. Fingerprints of global warming on wild animals and plants. Nature 421(6918):57-60.

Rosenzweig, C., D. Karoly, M. Vicarelli, P. Neofotis, Q. Wu, G. Casassa, A. Menzel, T.L. Root, N. Estrella, B. Seguin, P. Tryjanowski, C. Liu, S. Rawlins, and A. Imeson. 2008. Attributing physical and biological impacts to anthropogenic climate change. Nature 453(7193):353-357.

Sagarin, R.D., J.P. Barry, S.E. Gilman, and C.H. Baxter. 1999. Climate-related change in an intertidal community over short and long time scales. Ecological Monographs 69(4):465-490.

Santer, R.M. 2010. Benefits of a professional training year for undergraduates on a neuroscience degree scheme. Journal of Undergraduate Neuroscience Education 9(1):A1-A4.

Schimel, D., M. Keller, S. Berukoff, B. Kao, H. Loescher, H. Powell, T. Kampe, D. Moore, and W. Gram. 2011. NEON Science Strategy: Enabling Continental-Scale Ecological Forecasting. Available: http://www.neoninc.org/sites/default/files/NEON_Strategy_2011u2.pdf [accessed April 10, 2014].

Sharp, P.A., C.L. Cooney, M.A. Kastner, J. Lees, R. Sasisekharan, M.B. Yaffe, S.N. Bhatia, T.E. Jacks, D.A. Lauffenburger, R. Langer, P.T. Hammond, and M. Sur. 2011. The Third Revolution: The Convergence of the Life Sciences, Physical Sciences, and Engineering. Washington, DC: Massachusetts Institute of Technology. Available: http://dc.mit.edu/sites/dc.mit.edu/files/MIT%20White%20Paper%20on%20Convergence.pdf [accessed February 26, 2014].

Seymour, E., A.B. Hunter, S.L. Laursen, and T. DeAntoni. 2004. Establishing the benefits of research experiences for undergraduates in the sciences: First findings from a three-year study. Science Education 88(4):493-534.

Sorlin, S. 2012. Environmental humanities: Why should biologists interested in the environment take the humanities seriously? Bioscience 62(9):788-789.

United Kingdom (UK) House of Lords Science and Technology Committee. 2013. Scientific Infrastructure. Second Report. Available: http://www.publications.parliament.uk/pa/ld201314/ldselect/ldsctech/76/7602.htm [accessed April 17, 2014].

UNFCCC (United Nations Framework Convention on Climate Change). 2009. Copenhagen Accord. United Nations. FCCC/CP/2009/L.7. Available: http://unfccc.int/resource/docs/2009/cop15/eng/l07.pdf [accessed June 10, 2014]

USDA (U.S. Department of Agriculture). 2014. Gone but Not Forgotten: NWRC Closed Field Stations. Available: http://www.aphis.usda.gov/wildlife_damage/nwrc/about/history/Closed%20Field%20Stations2.pdf [accessed April 17, 2014].

van Alstyne, M. 1997. The state of network organization: A survey in three frameworks. Journal of Organizational Computing and Electronic Commerce 7(2/3):83-151.

Wagner, C. 2008. The New Invisible College: Science for Development. Washington, DC: Brookings Institution Press. 157pp.

Whitesell, S., R.J. Lilikeholm, and T.L. Sharik. 2002. A global survey of tropical biological field stations. BioScience 52(1):55-64.

Wilson, E.O. 1998. Consilience: The Unity of Knowledge. New York: Knopf.

Winker, K. and J.J. Withrow. 2013. Natural history: small collections make a big impact. Nature 493: 480. Available: http://www.nature.com/nature/journal/v493/n7433/full/493480b.html [accessed April 17, 2014].

Wyman, R.L., E. Wallensky, and M. Baine. 2009. The activities and importance of international field stations. Bioscience 59(7):584-592.

Zhang, D., B. Guo, and Z. Yu. 2011. The emergence of social and community intelligence. Computer 44(7):21-28. Available: http://www-public.itsudparis.eu/~zhang_da/pub/Daqing-IEEE-Computer-2011-07.pdf.

Appendix A
Statement of Task

A committee of the National Academies will conduct a study to review and assess the role of field stations, marine laboratories, and natural reserves (FSMLNRs) in science and engineering research, innovation, education, training, and public outreach and engagement. The study will evaluate FSMLNR effectiveness as individual entities and as collaborative networks to address local, national, or global challenges; their value as resources for environmental research; and provide suggestions for financially feasible approaches for the sustained operation and management in support of their often multifaceted roles. In particular the study will:

1. Assess the past and present contributions of FSMLNRs to
   a. Research and innovation to address pressing environmental and societal challenges.
   b. Education and training of the next generation of leaders in science and other disciplines.
   c. Public outreach mechanisms that enable individuals and communities to access, interpret, and use, or contribute to environmental science and engineering research.
2. Outline strategies for FSMLNRs to fill gaps in knowledge, open new avenues of inquiry (e.g., collaborations with industry), and forge a new convergence of science and engineering to advance research and innovation, education and training, and public outreach and engagement programs to form a new environmental infrastructure that can serve society at all levels.
3. Outline the infrastructure and logistical needs for FSMLNRs to fulfill their roles. Include perspective on physical (laboratories, research vessels, housing, transportation, canopy towers, class rooms etc.), technical (i.e., lab equipment, sensor arrays, etc.) and cyberinfrastructure needs to support or enhance their ability to benefit science and society. How can FSMLNRs be equipped to address and adapt to rapidly changing needs or capabilities in science and engineering, education, and public outreach and engagement?
4. Explore the potential for broader networking of FSMLNRs with other field facilities such as state and national parks and wildlife refuges among others.
5. Describe best practices and metrics that will enable FSMLNRs to monitor, assess, and modify their strategies to meet research and innovation, education and training, and outreach and engagement goals.

6. Suggest a range of long-term financial strategies that could be used for sustained support of FSMLNR individual and collective roles in research and innovation, education and training, and public outreach and engagement, including potential partnerships with industry to develop green technologies.

Appendix B
Committee Member Biographies

Jerry R. Schubel has been president and CEO of the Aquarium of the Pacific (AOP) in Long Beach, CA, since 2002. Before that, he was president and CEO of the New England Aquarium (1994–2001) and dean and director of the Marine Science Research Center of the State University of New York at Sony Brook (1974–1994). Throughout his professional life, Dr. Schubel has worked at the interfaces of science, management, and policy on ocean issues. He has published more than 225 scientific papers and has written extensively for general audiences. He is a member of the National Oceanic and Atmospheric Administration's Science Advisory Board, the Science Advisory Panel for California's Ocean Protection Council, and the Board of Governors of the Savannah Ocean Exchange. He chaired the National Sea Grant Review Panel, the National Research Council's Marine Board, and the Ocean Research and Resources Advisory Panel. He has served on numerous National Research Council committees, is a former member of the Environmental Protection Agency's Science Advisory Board, the Census of Marine Life U.S. National Committee, and the National Science Foundation's Education and Human Resources Advisory Committee. Dr. Schubel received an honorary doctorate from the Massachusetts Maritime Academy in 1998 and holds a Ph.D. in oceanography from Johns Hopkins University.

Felicia C. Coleman is the director of the Florida State University Coastal and Marine Laboratory. Dr. Coleman is a marine ecologist with a particular interest in reef fish behavior and use of habitat. She also focuses on how scientific findings are incorporated into laws and regulations that affect the management and conservation of living marine resources. Dr. Coleman has served on a number of committees and councils focused on conservation of marine resources including the Gulf of Mexico Fishery Management Council, the Marine Protected Areas Federal Advisory Committee, and the National Research Council.

Cathy Conrad is a professor in the Department of Geography of Saint Mary's University and adjunct professor at Dalhousie University and Wilfrid Laurier University (WLU) in Waterloo, Ontario. Her research encompasses fluvial geomorphology, watershed management, community-based environmental monitoring, and water quality. She has been involved with numerous environmental stewardship groups. She currently serves as research coordinator for the Community-Based Environmental Monitoring Network. She is actively involved in community-based conservation management projects in Cuba, Vietnam, and a number of sub-Saharan West African nations. Dr. Conrad received her B.A. (Honors, First Class) from Saint Mary's University in 1993, her master of environmental studies from Wilfrid Laurier University in 1995, and her Ph.D. in geography in the joint Waterloo-WLU graduate program in 2000.

Diane M. Debinski is a professor in the Department of Ecology, Evolution, and Organismal Biology of Iowa State University. She focuses her research on understanding and predicting species distribution and abundance patterns across the landscape on local and regional scales. Those patterns, when analyzed for spatial or temporal trajectories, can become bioindicators of climate change. In mountain systems, Dr. Debinski has studied the responses of plant and animal species to drought, warming conditions, and reduced snowpack. She has studied how landscape configuration, context, and management affect local and regional species patterns in prairie and grassland systems. Dr. Debinski received her B.A. from the University of Maryland in 1984, her M.S. from the University of Michigan in 1986, and her Ph.D. from Montana State University in 1991.

Peter M. Kareiva (NAS) is chief scientist of The Nature Conservatory, where he is responsible for developing and helping to implement science-based conservation throughout the organization. He joined The Nature Conservancy's staff in 2002 after more than 20 years in academe and work at the National Oceanic and Atmospheric Administration, where he directed the Northwest Fisheries Science Center Conservation Biology Division. In addition to his duties as the Conservancy's chief scientist, Dr. Kareiva's current projects emphasize the interplay of human land use and biodiversity, resilience in the face of global change, and marine conservation. He is a member of the National Academy of Sciences. Dr. Kareiva received an M.S. in environmental biology from the University of California, Irvine and a Ph.D. in ecology and evolutionary biology from Cornell University.

George I. Matsumoto has been the senior educational and research specialist at the Monterey Bay Aquarium Research Institute (MBARI) in Moss Landing, CA, since 1996. His research interests include pelagic and benthic communities, ecology, and biogeography of pelagic and benthic organisms, and functional morphology, natural history, and behavior of pelagic and benthic organisms. In addition to performing research at MBARI, Dr. Matsumoto manages several education and outreach efforts, including the seminar program, the internship program, and collaborations with MBARI's sister organization, the Monterey Bay Aquarium. Past professional experience includes teaching at Flinders University in Australia and serving as a National Science Foundation postdoctoral fellow. Dr. Matsumoto is an adjunct professor at Monterey Peninsula College. He received his Ph.D. in biological sciences from the University of California, Los Angeles in 1990.

Diane M. McKnight (NAE) is a professor of civil, environmental, and architectural engineering at the University of Colorado. Her research focuses on interactions between hydrological, chemical, and biological processes in the control of dynamics in aquatic ecosystems. That research is carried out through field-scale experiments, modeling, and laboratory characterization of natural substrates. Dr. McKnight also conducts research on interactions between freshwater biota, trace metals, and natural organic material in diverse freshwater environments, including lakes and streams in the Colorado Rocky Mountains and in the McMurdo Dry Valleys in Antarctica. She develops interactions with state and local groups involved in mine drainage and watershed issues in the Rocky Mountains. Dr.

McKnight is a member of the National Academy of Engineering. She is a former member of the National Research Council's Water Science and Technology Board and Polar Research Board. She received her Ph.D. in environmental engineering from the Massachusetts Institute of Technology in 1979.

Camille Parmesan is a professor of integrative biology at the University of Texas at Austin. She is also the National Marine Aquarium Chair in Public Understanding of Oceans and Human Health at the Plymouth University (UK) Marine Institute. Dr. Parmesan's research focuses on the current impacts of climate change on wildlife and ranges from field-based work on American and European butterflies to synthetic analyses of global impacts on a broad array of species in terrestrial and marine biomes. Dr. Parmesan collaborates with field stations in the United States, the United Kingdom, Spain, Finland, the Netherlands, France, and Australia for her work on the impacts of climate change on biodiversity. Through those collaborations, she has examined approaches to integrating databases to make research results on climate-change impacts available to the scientific community and for policy decisions. Dr. Parmesan works actively with government agencies and nongovernment organizations to help to develop conservation assessment and planning tools aimed at preserving biodiversity in the face of climate change. In 2007, she was awarded the Conservation Achievement Award in Science by the National Wildlife Federation, named Outstanding Woman Working on Climate Change by the International Union for Conservation of Nature, and included in *Who's Who of Women and the Environment* by the United Nations Environment Programme. Dr. Parmesan has been involved as an author and reviewer in multiple reports for the Intergovernmental Panel on Climate Change and is co-recipient with Al Gore of the Nobel Peace Prize awarded in 2007. She received her Ph.D. in zoology from the University of Texas in 1995.

Robert Plowes is a research scientist in the University of Texas Brackenridge Field Laboratory in Austin. His research focuses on understanding causes and consequences of biological invasions and the impacts of land use and land management on biodiversity. He studies host–parasite–pathogen interactions as a basis of biological control of invasive species, using molecular and microbial tools. He is also responsible for coordinating research and education activities and leading infrastructure development projects at two University of Texas field stations. Dr. Plowes received his Ph.D. in landscape ecology from the University of Texas at Austin in 2005 after a career as a consulting electric-systems engineer and engineering-business manager in Africa and the United States.

Alison G. Power is a professor in the Department of Ecology and Evolutionary Biology and the Department of Science and Technology Studies of Cornell University. At Cornell, she served as dean of the Graduate School in 2001–2010. Her research focuses on ecosystem services in agriculture, agroecology, interactions between agricultural and natural ecosystems, and disease ecology in plant communities. She is a past president of the Ecological Society of America and of the Association of Graduate Schools. She serves on the Board on Life Sciences of the National Academies, the U.S. National Committee for DIVERSITAS, and the Great Lakes Bioenergy Research Center Scientific Advisory Board. Dr. Power received a B.S. in biology from the University of Alaska Fairbanks and a Ph.D. in zoology from the University of Washington.

Mary Power (NAS) is a professor of integrative biology at the University of California, Berkeley and is the faculty manager of Angelo Coast Range Reserve. Her research interests center on river food webs and the interactions among fish, birds, invertebrates, and algae in temperate and tropical rivers. Dr. Power is especially interested in how attributes of species affect food-web structure and dynamics and how the strengths of the interactions change under different environmental regimes. Much of Dr. Power's field work takes place in the South Fork Eel River in the Angelo Coast Range Reserve in Mendocino, CA, one of the University of California Natural Reserve System's 35 research and teaching reserves. Dr. Power received her B.A. from Brown University in 1971, her M.S. from the Boston University Marine Program in 1974, and her Ph.D. from the University of Washington in 1981.

Mark R. Stromberg was from 1988 to 2011 the resident director of the Hastings Natural History Reservation, a reserve in the University of California (UC) Natural Reserve System (NRS), established in 1937 by the Museum of Vertebrate Zoology of UC, Berkeley for advanced research and teaching in field biology. In 2011, he moved to a position with the UC NRS in the UC Office of the President. At Hastings, Dr. Stromberg coordinated all the research on the reserve; managed the facility maintenance; developed, administered, and maintained the reserve's computer network; served as data manager and Web manager; planned long-term projects; hosted visiting groups; represented the reserve to local and regional organizations and government agencies; represented the reserve in national and regional organizations (such as CalEON, the California Biodiversity Center, the Organization of Biological Field Stations, and the UC NRS); oversaw safety and animal-care issues; and functioned as co-principal investigator on grants. Dr. Stromberg managed the first of many California Proposition 84 grants to the UC NRS to install new windows, electric service, plumbing, and insulation and many other needed upgrades in the older buildings. He arranged over $4 million in funding for infrastructure at Hastings, including new laboratories, classrooms, barns, garages, and housing for up to 40 visitors. Developing other funding, Dr. Stromberg collaborated with the Western Regional Climate Center to install online weather stations at Hastings and 18 other UC reserves. Developing funding from the American Recovery and Reinvestment Act (ARRA), he coordinated the installation of a fast radio link to the Internet and provided wireless Internet access essentially anywhere on Hastings. He also coordinated similar cyberinfrastructure installations in 14 other NRS reserves with ARRA funds. Dr. Stromberg is assisting in writing a strategic plan for the NRS to focus on seven themes to develop the strengths of NRS as a network over the next 10 years. He received a B.S. in wildlife biology from Colorado State University in 1973, an M.S. in zoology from the University of Wisconsin–Madison in 1975, and a Ph.D. in zoology in 1979 from the University of Wisconsin–Madison.